# TRANSPORTING EPITHELIA

*MICHAEL J. BERRIDGE*

Agricultural Research Council
Unit of Invertebrate Chemistry and Physiology
Department of Zoology
University of Cambridge
Cambridge, England

*JAMES L. OSCHMAN*

Department of Biological Sciences
Northwestern University
Evanston, Illinois

*ACADEMIC PRESS* New York and London

ACADEMIC PRESS, INC.
111 Fifth Avenue, New York, New York 10003

*United Kingdom Edition published by*
ACADEMIC PRESS, INC. (LONDON) LTD.
24/28 Oval Road, London NW1 7DD

LIBRARY OF CONGRESS CATALOG CARD NUMBER: 74-187244

PRINTED IN THE UNITED STATES OF AMERICA

# CONTENTS

# ABBREVIATIONS

| | | | | |
|---|---|---|---|---|
| af | apical fold | lpm | lateral plasma membrane |
| apm | apical plasma membrane | ly | lysosome |
| az | adhering zonule | m | mitochondrion |
| bf | basal fold | mc | myoepithelial cell |
| bi | basal interdigitation | mv | microvilli |
| bm | basement membrane | n | nucleus |
| bpm | basal plasma membrane | om | occluding macula—gap junction |
| cap | capillary | oz | occluding zonule—tight junction |
| d | desmosome-adhering macula | pg | protein granule |
| e | erythrocyte | pm | plasma membrane |
| G | Golgi complex | rer | rough endoplasmic reticulum |
| g | secretory granule | sd | septate desmosome |
| is | intercellular space | ser | smooth endoplasmic reticulum |
| jc | junctional complex | v | vacuole |
| L | lumen | | |

## PREFACE

Early studies of transporting epithelia used the black box approach. Substances were added to one side of an epithelium, and the rates at which they appeared on the opposite side were measured. Studies of this kind were invaluable in defining the biophysical parameters of transporting systems; but in order to learn more about the mechanism of transport it was necessary to open the box and study its contents. Although the surfaces responsible for transepithelial movement of charged molecules were located with microelectrodes, and biochemists found that the energy for transport comes from membrane-bound adenosinetriphosphatases, the electron microscope has been the single most important technique to open the transport black box. Electron microscopy has defined the structural organization of each epithelium so that it can now be integrated with its function to yield a detailed concept of the transport mechanism.

This monograph summarizes the progress that has been made in understanding a wide range of epithelial transport systems. We discuss epithelia involved in osmotic and ionic regulation from protonephridia to the mammalian kidney, digestive and absorptive epithelia, and epithelia that produce special secretions such as milk, endolymph, aqueous humor, cerebrospinal fluid, sweat, and tears.

We have not attempted a comprehensive review of the immense literature in the fields of physiology and ultrastructure which continues to grow at an increasing rate. Instead we use a similar format to summarize the essential features of each epithelium. First, the role of the epithelium in the physiology of the animal is presented. Second, its structure is described with the aid of detailed diagrams and electron micrographs. Finally, the structure of the epithelium is correlated with its physiological properties, sometimes for the first time. At times our understanding of structure or physiology is inadequate, and we hope that this volume may stimulate others to close the gaps in our knowledge. We also hope our approach makes the monograph valuable both for teaching and as a reference for research workers interested in comparative aspects of transport phenomena.

MICHAEL J. BERRIDGE
JAMES L. OSCHMAN

## ACKNOWLEDGMENTS

We are greatly indebted to Betty J. Wall and Brij L. Gupta for their invaluable help and advice during the preparation of this volume. We thank Albert Cass and Alexander Mauro for their comments on portions of the manuscript. Bonnie J. Sedlak gave expert assistance with the preparation of illustrations and Margaret Clements with the typing of the references. We thank librarians Ruth Spiecker and Mary Ann Barragry for checking references. We wish to express our debt to those who contributed the electron micrographs and gave permission to use their published and unpublished pictures. We also thank the various publishers for permission to reproduce published material.

## INTRODUCTION

Animals are comprised of epithelial sheets and tubes separating fluids of different compositions. Molecules can be moved across epithelia although the individual cells continue to maintain the constant internal environment characteristic of all cells. Net transport across epithelia is accomplished because their constituent cells are structurally and functionally asymmetric—the membranes facing the two surfaces differ in both their geometry and in their chemical properties. During the past 10 years collaboration among anatomists, physiologists, and biophysicists has led to a concept that may have general application to transporting epithelia. The osmotic gradients that produce transepithelial water movements are created by pumping solutes into confined compartments within epithelia. These compartments are most likely located between folds of the cell surfaces. This folding, which takes various forms, is the single common anatomic feature of transporting epithelial cells. The widespread applicability of this new concept has changed the way we think about the functional architecture of transporting cells and has stimulated the preparation of this volume.

## TRANSFER ACROSS MEMBRANES

The interface between the cell interior and the environment is a plasma membrane that appears in electron micrographs of profiles as two dense lines separated by a less dense region. Physiologists have been concerned with the possible physical mechanisms by which molecules move across membranes. The plasma membrane consists of an organized mixture of polar lipids, carbohydrates, and proteins. The lipids form a bilayer structure that is exceedingly impermeable to all but lipid-soluble substances. The latter can traverse the membrane by diffusion (Fig. 1 a). Since lipid-insoluble molecules such as electrolytes and carbohydrates traverse biological membranes, substances other than lipids must be present to modify the bilayer structure to allow the insoluble molecules to pass through. Many sorts of molecules may cross membranes by combining with protein carriers that can move across the membrane. Transport involving carriers (Fig. 1 b) can be passive (facilitated and exchange diffusion) or active (active transport). Lipid-insoluble substances may move through holes or pores (Fig. 1 c). This type of movement depends on the particle size and

shape relative to the pore diameter, and on the charges on the molecule and pore.

## WATER MOVEMENT ACROSS EPITHELIA

A number of explanations have been considered for the movement of water across epithelia, including classical osmosis, active transport of water, and pinocytosis. The main argument against classical osmosis is that water traverses many epithelia in isosmotic proportions in the absence of a net osmotic gradient across the epithelium. Examples include isosmotic secretion of HCl by the stomach, of bile by the liver, of cerebrospinal fluid by the choroid plexus, pancreatic juice by the pancreas, and urine by the insect Malpighian tubule. Further, isosmotic reabsorption occurs in the intestine, gallbladder, and kidney proximal tubule. In some cases slight transepithelial osmotic gradients have been measured, but they have been insufficient to generate the rates of fluid flow that are actually observed. Water movement against an osmotic gradient occurs in the insect rectum and in *in vitro* preparations of gallbladder and intestine. In the latter three examples, water traverses the membrane in a direction opposite to that expected from osmosis. Active transport of water has been considered as a mechanism for moving water against an osmotic gradient. However, this view is no longer favored for most epithelia because water movement is usually closely linked with solute transport. Pinocytosis is an effective method for moving molecules across membranes through the formation of vesicles by invagination of the cell surface. Pinocytotic vesicles are observed frequently in transporting cells, where they afford a mechanism of exchange of water and solutes between the exterior and isolated compartments within the cells (endocytosis). Vesicular transport cannot explain fluid transfer across epithelia, mainly because it is inconsistent with the high degree of specificity observed when many different compounds are tested for their ability to cross epithelia. Pinocytosis seems to be important only in endothelia, such as corneal "endothelium" (132) and certain capillaries (26) which have functional vesicular transport systems that shuttle fluid from one side to another.

An important step in describing water movement across epithelia was the development by Curran and his colleagues (49–51, 190) of a three-compartment, double-membrane model to explain intestinal absorption. In this model, illustrated in Fig. 2 a, there are two membranes with different permeabilities separating

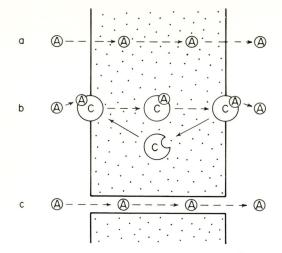

**Fig. 1.** Movement of molecules across membranes. a, Molecules soluble in the membrane can diffuse through it. b, Carrier mediated transport of a substance A by a molecule C within the membrane. A is bound to the carrier on one side of the membrane, moved across the membrane, and released from the carrier into the solution on the other side. The carrier then returns to the first side. When the reaction between A and C is passive, the mechanism is facilitated diffusion. When the carrier cannot return to side 1 without carrying another A molecule with it, the mechanism is exchange diffusion. If metabolic energy is involved in the formation or breaking of the bond between A and C, the mechanism may be active transport. c, Substances may also traverse the membrane by diffusing through pores. The rate of this process is determined by the pore shape, size, and charge relative to those of the diffusing molecule.

three compartments. The first membrane, 1, is impermeable to solutes, while the second membrane, 2, is porous and permeable to both water and solute. If solute is actively transported across membrane 1 from compartment L to compartment M, the concentration in compartment M will be elevated and water will enter from L by osmosis across semipermeable membrane 1. Since water is essentially an incompressible fluid, water movement into compartment M will produce a hydrostatic pressure against its walls. If the walls are inelastic, the hydrostatic pressure will force water and solute across the porous membrane 2. Thus a transport of solute across membrane 1 will produce a net fluid movement across the whole system from compartment L to M to R.

Kaye *et al.* (*136*) examined the ultrastructure of the gallbladder and found that the intercellular spaces

distend during transport and collapse when transport is stopped. They suggested that the intercellular spaces may correspond to compartment M of Curran's model (Fig. 2 b). Solute transport from the cell into the intercellular spaces (corresponding to compartment M of the model) would bring about an osmotic water flow in the same direction, and create a hydrostatic pressure against the lateral membranes. This would bring about distension of the spaces and create a hydraulic flow through the spaces and into the connective tissue and capillaries, corresponding to compartment R. The narrow basal slit and epithelial and capillary endothelial basement membranes would comprise membrane 2.

Diamond and his colleagues (*56–59, 277*) further refined this model (Fig. 2 c) by suggesting that there is a standing osmotic gradient within the intercellular spaces, with the closed apical end of the intercellular channel much more concentrated than the basal end. If the geometry of the channel, permeability of the channel membranes, and rate of solute transport and diffusion are appropriate, osmotic equilibration will occur progressively along the length of the channel so that the fluid emerging from the open end will be isosmotic to the fluid bathing the apical cell surface, and net isosmotic fluid transport will occur. Diamond and Bossert (*56–57*) developed a mathematical model incorporating the basic parameters expected to affect the concentration of the fluid emerging from the open end of a channel containing a standing gradient. The calculations revealed that any changes in the parameters that would reduce the opportunity for solute molecules within the channel to draw water from the cell would tend to produce a hyperosmotic absorbate. Examples of such changes would be decreasing channel length, increasing channel width, increasing the solute diffusion constant, and decreasing the water permeability of the channel wall. The relationships between the various parameters serve as useful guidelines for applying the standing gradient model to epithelia other than gallbladder.

The general model for a channel with fluid emerging from its open end is illustrated in Fig. 3 a. This is termed a *forward channel*, and is exemplified by intercellular spaces of absorptive epithelia, secretory canaliculi, spaces between microvilli of secretory epithelia, basal infolds of absorptive epithelia such as salivary gland striated duct and proximal and distal tubules of kidney. Many epithelia have channels for which the direction of flow is into the open end rather than out of it. Figure 3 b shows how the standing gradient model might work in reverse in such channels. In this *backward channel* system solute is pumped from the channel into the cell, diluting the channel contents. Examples include intercellular spaces of reptilian salt gland, microvilli and apical foldings of absorptive epithelia, and basal infolds and interdigitations of secretory epithelia. A favorable gradient is estab-

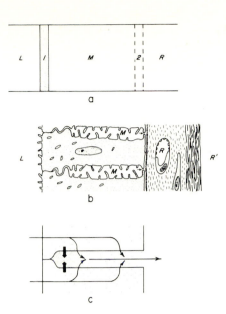

Fig. 2. Models used to explain water uptake by gall-bladder. a, Curran's double membrane model (49–51). The first barrier, 1, is impermeable to solutes in the model, and in a living membrane would correspond to the site of solute transport from L to M. The second barrier, 2, is a porous nonselective membrane. Solute is placed in the middle compartment (M) as by active transport from the left compartment (L). This creates an osmotic gradient across semipermeable barrier 1 so that water moves from L to M. This elevates the hydrostatic pressure in compartment M, driving fluid from M to R via the porous membrane 2. b, General structure of the gallbladder epithelium showing the analogy drawn by Kaye *et al.* (136) to Curran's model. The middle compartment (M) is the intercellular space. Solute transport into this compartment elevates the osmotic pressure there, drawing water in and distending the interspace. Fluid then flows out through the narrow opening at the basal surface, through the epithelial basement membrane, connective tissue space, and endothelial basement membrane into the capillaries. The latter constitute the compartment R of the model. When the gallbladder is functioning *in vitro*, the serosal bath (R') becomes the third compartment. c, The standing gradient model developed by Diamond and his colleagues (55–59). The closed end of the intercellular channel is made considerably hyperosmotic to the cell by the action of solute pumps. Water flows into the channel to equalize the osmotic gradient, so that isosmotic fluid emerges from the open end.

lished for movement of water from the channel into the cell. If the channel is rigid and will not collapse, water entering the cell will be replaced by fluid flowing into the open end of the channel. Equilibration in this system is affected by the same parameters as in the forward channel system.

The possibility of widespread application of the standing gradient model is apparent when it is realized that the main characteristic of transporting cells is extensive folding of the cell surfaces, forming long and narrow channels that might be analogous to the intercellular spaces of gallbladder. The various types of surface foldings and examples of them are illustrated in Fig. 4. It is frequently stated that such folding provides a greater surface area for diffusion than would a flat membrane. Although this is true, these foldings could have a more profound geometric significance as the sites of standing gradients. The general applicability of the standing gradient hypothesis provides a central

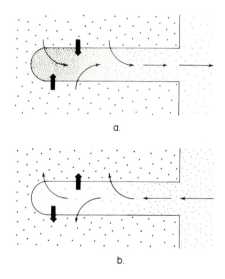

Fig. 3. Standing gradient models. a, The forward channel, in which fluid flows out from the open end owing to the standing gradient created within the space by active transport of solute from the cell. Examples include intercellular spaces of absorptive epithelia, secretory canaliculi, spaces between microvilli of secretory epithelia, basal infolds of absorptive epithelia such as salivary gland striated duct and proximal and distal tubules of kidney. b, Backward channel, with fluid entering the open end owing to the reverse standing gradient created by solute uptake into the cell. Examples include intercellular spaces of reptilian salt gland, microvilli and apical foldings of absorptive epithelia, and basal infolds and interdigitations of secretory epithelia.

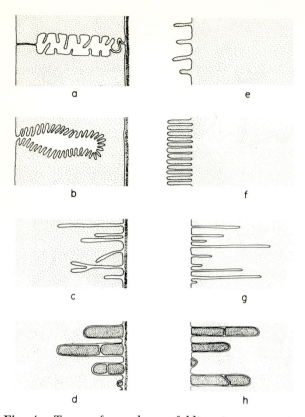

**Fig. 4.** Types of membrane folding in transporting epithelia. a, Intercellular space (gallbladder, intestine, rat proximal tubule, frog skin, insect rectum, toad bladder, rumen, reptilian salt gland, amphibian parietal cell). b, Canaliculus (chloride cell, insect Malpighian tubule, vertebrate salivary gland, insect salivary gland, mammalian parietal cell, liver, pancreas, insect goblet cell). c, Basal infolds (Malpighian tubule, avian salt gland, choroid plexus). d, Basal interdigitations (salivary gland striated duct, kidney proximal and distal tubules, ciliary body, segmental organs, some Malpighian tubules). e, Microvilli type I, short and widely spaced (gallbladder, toad bladder, salt gland, rectal gland, pancreas, liver, mammalian salivary gland, parietal cell, mammary gland, lacrimal gland, stria vascularis). f, Microvilli type II, long and closely packed (intestine, kidney proximal and distal tubule, insect Malpighian tubule, coxal gland, green gland, insect midgut goblet cell). g, Apical infolds (anal papillae, crustacean gill, insect rectum). h, Apical interdigitations (kidney proximal tubule, rectal gland).

theme for the survey of transporting epithelia presented in this volume.

It must be recognized that the standing gradient model represents a highly plausible hypothesis which

has yet to receive extensive testing. Attempts have been made to verify the hypothesis. An indirect approach has been to determine if the various channels thought to be involved in standing gradient flow distend during transport and collapse when transport is inhibited, as was demonstrated for gallbladder (*136, 277*). This phenomenon has been confirmed for some epithelia but not for others. It must be realized that a channel does not have to distend when fluid is flowing through it. Cells possess many structural elements, including tonofilaments, desmosomes, and microtubules, that serve to maintain a constant form when stresses are placed upon them, such as the hydrostatic pressure generated by osmotic flow into a narrow compartment. Also, changes in cell volume occur during preparation of tissues for histological examination, so much caution is required when using this method for comparing the sizes of channels. Direct observation on living tissues is more reliable, as has been accomplished with the isolated collecting duct of kidney (*96*), toad bladder (*97*), and insect rectum (*297*). In the latter example, micropuncture samples from the intercellular spaces consistently had a higher osmotic pressure than the fluid in the rectal lumen during fluid uptake (*297*). This evidence, along with estimates made from measurements of streaming potentials in gallbladder (*174*) confirms that gradients are developed within intercellular spaces during fluid transport.

There is a striking similarity between the physiological characteristics of isosmotic fluid absorption by epithelia such as gallbladder and isosmotic secretion by epithelia such as insect Malpighian tubules and salivary glands (*198*). If it is accepted that standing gradients within intercellular spaces bring about fluid uptake in absorptive epithelia, where are these gradients formed in secretory epithelia which lack extensive intercellular spaces, but which have instead elaborate basal infoldings and microvilli? Perhaps the gradients are located within the infoldings and between microvilli. Such channels are shorter than intercellular spaces and may not be long enough to sustain large standing gradients. Perhaps the gradient between the two microvilli is minute but the cumulative effect of the thousands of microvilli on the surface of each secretory cell is sufficient to bring about isosmotic fluid transport. The gradients within basal infoldings and between microvilli have not been measured.

Elaborate surface folding is often most pronounced at only one side of the cell, as in gallbladder, salt gland, rectal gland, and stria vascularis. We have seen how foldings of the cell surface can provide sites for local osmotic gradients coupling ion transport to water flow, but we must still explain how fluid traverses the opposite membrane of those cells which have the folding confined to one side. While this problem has not been given much attention, it does appear that both artificial and natural membranes have sufficient water permeability so that fluid can cross them at a reasonable rate. The rate of water flow across a semipermeable membrane is identical when produced by a given number of atmospheres of osmotic pressure difference or the same number of atmospheres of hydrostatic pressure. Fluid transport out of a cell would tend to shrink it, but structural rigidity of the cell together with a high water permeability of the plasma membrane may allow for hydraulic flow across one surface to replace fluid pumped out across the other. Similarly, fluid transport into a cell would tend to make it swell, but this would also force fluid out across the opposite surface.

## SOLUTE–SOLVENT INTERACTIONS IN EPITHELIA

The existence of folded membranes at the surfaces of epithelial cells has a number of consequences for the flow of nontransported solutes. Physiologists have begun to explore this problem, but much more detailed information will be required concerning the interactions between solutes and solvent occurring within long and narrow channel systems. Most of the following models have been devised by physiologists in an effort to explain a particular phenomenon they have observed. They serve to illustrate some of the complexity which can arise from the presence of extracellular compartments at the surfaces of epithelia.

### Sweeping-in Effect

Water movement across a membrane can carry solutes by solvent drag (3). Suppose a membrane has pores through which water and solute move. Solute flux will be higher in the direction of water flow, as in this direction the rate of water flow is superimposed upon the rate of solute diffusion. The solute flux will be correspondingly decreased in the opposite direction. Gradients in concentration of the solute will develop adjacent to the cell surface since water movement through the membrane will cause a piling up of solute on one side and dilution or washing away on the opposite side. Thus water movement will produce a concentration gradient favoring solute movement in the same direction. Now consider the backward channel system shown in Fig. 3 b. Solutes in the bathing fluid will be swept into the channel by the bulk flow of fluid. If the channel is very deep and narrow, the solvent drag effect will be much greater than for a flat membrane, i.e., once a solute has been swept into the channel, diffusion back out will be hindered by the inward flow. Any solutes accumulating at the base of the channel will tend to diffuse across the membrane into the cell.

### Osmotic Filtration

The effect just described may be important in many epithelia that produce fluids similar in composition to a filtrate of the blood, i.e., containing the same ions (although in somewhat different proportions) and lacking proteins. Filtration can often be ruled out because the hydrostatic pressure developed by fluid secretion, as measured at the ducts, greatly exceeds the hydrostatic pressure of the blood supply to the secretory organ. Examples include the salivary glands, choroid plexus, aglomerular kidney, and insect Malpighian tubule. Osmotic filtration (Fig. 5) could provide a simple explanation for fluid secretion in these systems (198). Here fluid is drawn through a filter, such as the epithelial basement membrane or underlying lamina propria, as a consequence of solute-linked water uptake from a basal compartment such as a basal infold. Nontransported solutes that are able to penetrate through the filter may accumulate within the channel, creating a gradient that augments their diffusion across the membrane.

Osmotic filtration does not seem to be applicable to forward channels such as intercellular spaces, since the cell and tight junction or occluding zonule (see below) are interposed between the source of the fluid and the channel, and it has always been assumed that the occluding zonule was an effective barrier to transport from the bathing medium into the channel. However, it is now clear that the occluding zonules are leaky in at least some epithelia (see below), so that some water and solutes may flow through the occluding zonule rather than through the cell. For this reason osmotic filtration might occur at the level of the occluding zonule.

### Entrainment

Entrainment is an effect similar to solvent drag that may operate in forward channels. In many epithelia the forward channels are organized so that there is a large isolated extracellular compartment with a restricted access to the blood. This is particularly evident in intestine (p. 56) and gallbladder (p. 52) of vertebrates, in insect rectum (p. 20) in which the intercellular spaces are extensive, and in insect intestine (p. 58) in which the basal infoldings form a compartment with only a few openings into the blood. The geometry of the cell surface in these epithelia is such that molecules absorbed by the epithelia must pass through the enclosed space (Fig. 6). If ions (e.g., $Na^+$) and water are transported into this space, it will have a composition that is different from that on either side of the epithelium, owing to the limited communication between the space and the blood. Since the compartment is being continually flushed out with fluid from the cell, the compartment will have the lowest concentra-

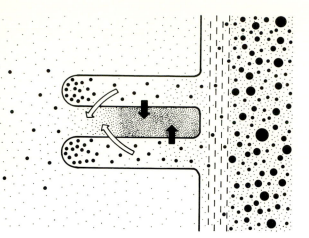

**Fig. 5.** Osmotic filtration. Solute transport from the basal channels (solid arrows) creates a flow of water into the cell (open arrows) as in the backwards channel system of Fig. 3. Bulk fluid flow carries solutes into the channel to the extent that they can penetrate the basement membrane, while protein molecules are left behind. Solutes that do not easily penetrate the plasma membrane would pile up towards the closed ends of the basal channels, establishing a more favorable gradient for their entry into the cell.

tion of a nontransported substance, X, such as an organic molecule. Continuous flow of fluid through the narrow exit channel will prevent X from diffusing back from the blood into the extracellular compartment. This will create a gradient favoring diffusion of X from the cell into the compartment which in turn lowers the concentration of X in the cell, establishing a gradient for X to enter either from the lumen or from the blood. If the plasma membrane facing the blood has a much lower permeability to X than does the luminal surface, there will be a net movement of X from lumen to blood against a concentration gradient. Uptake of solute in this manner has been termed *entrainment* (*14*). It should be noted that if water is obtained from the lumen side, the mechanism is an elaboration of solvent drag, in which the flow of water is superimposed on the rate of solute diffusion. However, the system could also work if water were drawn from the basal side. In this case fluid flow through the extracellular compartment would provide a gradient for diffusion of the entrained solute, X, from the lumen. In this case X would move across the epithelium in the absence of a net fluid uptake in the same direction. A mechanism of this sort may aid in the uptake of certain organic molecules by insect midgut and rectum and might also account for urea uptake by kidney collecting ducts (p. 38).

## Solute Recycling

Another modification of the system described above is for the actively transported solute to be obtained from the blood surface rather than from the lumen (Fig. 7). This mechanism may explain the operation of the insect rectum, which is able to absorb water even when there are no transportable solutes in the lumen (*210*). Suppose standing gradients are produced by solute transport into the intercellular spaces. This would tend to deplete the cell of solute. Since solute cannot be replaced from the lumen, solutes may instead be recruited from the blood. Alternatively, solutes could be recycled within the epithelium (*17, 199, 295, 296*). In either case the net effect is the same, i.e., each solute molecule can be used many times to contribute to the standing gradient that draws water from the lumen. This results in a net hypoosmotic uptake from the lumen, i.e., water molecules move in excess of solute from lumen to blood.

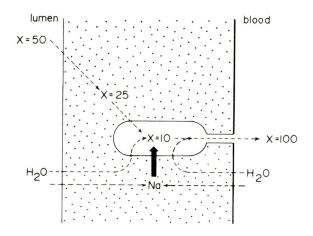

**Fig. 6.** Entrainment. A solute, X, moves from lumen to blood against a concentration gradient without being actively transported. The movement of X is determined by the diffusion gradient between the lumen and the extracellular compartment (EC) which corresponds to the intercellular spaces of absorptive epithelia such as vertebrate intestine or the basal channel system of insect midgut. Water moves into EC as a result of active sodium transport, and maintains a favorable gradient for the passive influx of X by continuously flushing out EC. The flow of X is entrained in the flow of fluid. The restricted opening into the blood favors the outward flow while reducing the diffusion of X from blood to EC. The water and sodium might be drawn from either side of the epithelium.

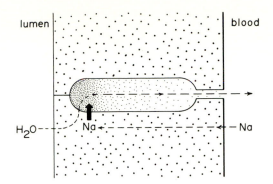

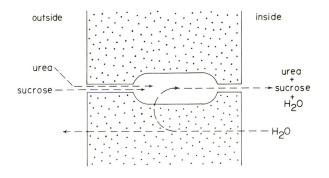

from the outside to the inside, and sucrose will be carried inward by the anomalous solvent drag so that the net flux of sucrose from outside to inside will be several times the flux in the opposite direction. The chemical potential difference produced by the urea gradient is the sole energy source for the movement of sucrose.

**Fig. 7.** Solute recycling. Studies of the insect rectum show that water can be taken up from the lumen in the absence of net solute uptake. Solutes recruited from the blood or recycled within the epithelium maintain the standing gradients within the intercellular spaces. This model could be combined with entrainment (Fig. 6) to maintain uptake of a number of solutes such as amino acids and sugars in the absence of a net flux of ions from lumen to blood.

## Anomalous Solvent Drag

Ussing (288) has described another application of entrainment. Known as anomalous solvent drag, this mechanism can operate to move solute against a gradient in the absence of active solute transport. The outside of a frog skin is made hyperosmotic to the inside by addition of urea or some other small molecule (Fig. 8). The urea appears to cause an opening of the tight seals or occluding zonules. The evidence for this is that the water permeability increases, the electrical resistance drops to a very low level, a nonselective shunt path is created for sodium chloride, molecules such as sucrose that do not readily enter cells traverse the skin more easily, and tracers such as Ba and $SO_4$ added to opposite sides of the membrane meet and precipitate within the interspaces (288–290). Opening of the occluding zonule provides a pathway so that urea can diffuse into the space between the cells. This elevates the osmotic pressure in the interspace, and water is drawn from the cells. Water entering the interspace tends to flow toward the inside surface of the skin because the mucous coat on the outside hinders outward hydraulic flow (although it allows inward diffusion of small molecules). The water movement toward the inside of the skin carries with it urea and any other small molecules that have diffused into the interspace from the outside fluid. Suppose, for example, that the outside is made hyperosmotic to the inside by addition of 200 mmoles/liter of urea, and that 10 mmoles/liter sucrose is present on each side of the skin. Urea will diffuse down its gradient

**Fig. 8.** Anomalous solvent drag. Ussing (288) has described this situation in which an osmotic gradient across an epithelium is the sole source of energy that brings about solute uptake in the direction opposite to osmotic water flow. The outside of a frog skin is made hyperosmotic to the inside by adding urea or some other small molecule (for example, at a concentration of 200 mmoles/liter). The urea appears to cause an opening of the tight seal or occluding zonule, and diffuses via that route into the space between the cells. The mucous coat on the outside of the skin allows inward diffusion of small molecules while hindering hydraulic flow in either direction. Urea entering the interspace elevates the osmotic pressure there, drawing water from the cells. Fluid tends to flow towards the inside of the skin rather than toward the outside, as there is less resistance to flow at the basement membrane. The fluid flow from the interspaces to the inside carries with it urea as well as any other small molecules that happen to diffuse into the interspace. For example, if 10 mM of sucrose is placed on both sides of the skin, the influx of sucrose (outside to inside) will be several times greater than the outflux. Water moving from the cell into the interspace is replaced by water entry into the cell across the basal surface, which is less resistant to water flow than the apical surface.

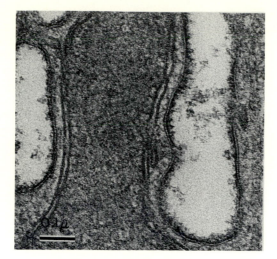

**Fig. 9.** Asymmetric plasma membranes are found in a number of transporting cells. In this example, from the reabsorptive segment of insect salivary gland (the basal surface, see p. 46), the outer leaflet of the membrane is coated with an electron dense material, the glycocalyx. Similar membrane asymmetry is found in urinary bladder, cornea, frog skin, intestine and others. ×100,000.

## STRUCTURAL CHARACTERISTICS OF TRANSPORTING CELLS

In addition to their characteristic folding of the cell surfaces, transporting epithelia have other structural features that may be more or less similar to those of nontransporting cells. Some of these will be described here.

### Membrane Asymmetry

It is known that cells are asymmetric—one surface may have a higher permeability to water or solutes than the other, or the transport of a solute may be confined to one surface. These physiological differences are sometimes reflected in differences in membrane folding and in differences in the appearance of the membranes themselves. For example, one leaflet of the plasma membrane may be thicker than the other. When the outer layer is thicker (as shown in Fig. 9) there may be an extracellular coating or glycocalyx (11). This material, commonly glycoprotein or mucopolysaccharide, forms the outermost interface between the cell and its environment, and has unusual properties with regard to charge distribution and water and ion binding capacity.

Another example is found in many transporting epithelia of insects, in which there are particulate subunits coating the cytoplasmic sides of the apical plasma membranes (Fig. 10). This was first observed in the rectal papillae of blowflies (100) and has subsequently been found in insect Malpighian tubules (19), salivary glands (197), rectal pads (199), and midgut goblet cells (4). The particles may be multienzyme complexes, perhaps including carbonic anhydrase and possibly ATPase. This is suggested because carbonic anhydrase of insects is in a particulate form (284), whereas it is soluble in vertebrates. Both carbonic anhydrase and the particles have been found in abundance in midgut goblet cells (4, 284).

### Junctions

Adhesion is the fundamental property of epithelial cells that enables them to form continuous sheets. Junctions serve to hold the cells together as well as to provide routes of communication from cell to cell. Junctions also act as permeability barriers to transepithelial fluid movement.

The most common junctions found in vertebrate epithelia are illustrated in Fig. 11. There is much variation in the position and number of the various sorts of junctions along the intercellular space. In many epithelia the juxtaluminal region has a terminal bar, the term used by light microscopists to describe the dark-staining region between adjacent columnar epithelial cells near their free surfaces. The terminal bar forms a continuous dense band around the perimeter of each cell. Electron microscopy reveals the terminal bar as a junctional complex usually consisting of an occluding zonule (tight junction), adhering zonule, and a desmosome (adhering macula). Additional desmosomes and gap junctions (occluding maculae) may also be found at other points along the length of the intercellular space.

The outermost occluding zonule or tight junction, as the name implies, is thought to form a continuous seal or barrier around the cell, acting as a hindrance to diffusion along the extracellular space. This interpretation is based on the observation that the outer leaflets of the adjacent plasma membranes appear to fuse (83, 94). Tracer molecules such as hemoglobin, peroxidase, and lanthanum fail to penetrate beyond the occluding zonule. However, in some epithelia the occluding zonule is not always closed to smaller molecules such as water and ions, and there is a definite extracellular leak or "shunt" between the cells. Mention has already been made of this, as the junctions appear to open during anomalous solvent drag across frog skin. Leaks may also be present under normal conditions in a number of epithelia, such as gallbladder (60) and kidney proximal tubule (92). Indeed, evidence is accumulating to suggest that the so-called tight junction is the principal route of passive ion fluxes across a number of epithelia (91).

The adhering zonule, like the occluding zonule, also forms an apical collar around the cell, but unlike the occluding zonule, it does not seem to restrict diffusion.

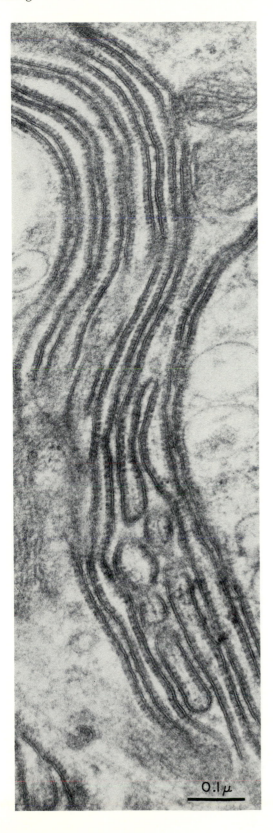

0.1 μ

The two unit membranes of the apposing cells remain approximately 200 Å apart and there is a homogeneous material in the interspace (Fig. 11). A belt of dense material is associated with the inner leaflets of each membrane. Fine fibers from the cytoplasm insert obliquely into the dense material. The adhering zonule is particularly evident in intestinal epithelia where it represents the point of attachment of the terminal web.

The desmosome (adhering macula) has a structural role in holding the cells together as would a spot weld between sheets of metal (Fig. 11). At these oval junctions the intercellular space can be increased to 300–400 Å and is filled with a homogeneous material bisected by a lamina of high density. A cytoplasmic plaque, considerably denser than that in the adhering zonule, is closely applied to the inner leaflet of the unit membrane. Long fibrils attached to each plaque pass out perpendicularly into the adjacent cytoplasm (137).

True occluding zonules are distinguished from the gap junctions which may occur at various sites along the intercellular cleft (Fig. 11). These are regions where the outer leaflets of the apposed membranes approach within 20 Å of one another to form a very narrow channel. These are maculae or spots rather than continuous bands around the cells. Thus they do not present a barrier to diffusion as molecules can move either through the narrow gap or can bypass it by going around the junctions (25). The 20 Å intercellular gap contains a closely packed array of particles with a repeating periodicity of about 90 Å (38, 226). Such gap junctions which occur in the transporting epithelia of kidney (256), liver (38), Malpighian tubule (19), and insect salivary gland (197), closely resemble the electrotonic junctions which represent low resistance pathways between cells in the brain and heart (25, 173, 226, 230). Apart from an obvious function in cell adhesion, gap junctions may be the sites where individual epithelial cells communicate with each other, thus enabling them to function in concert (169). The occluding zonule has been suggested as another possible low-resistance pathway between cells (168).

Epithelia of invertebrates have the usual junctions just described, but, in addition, have as the most prominent junctional element the septate desmosome (Fig. 12). These occupy a large portion of the intercellular channel and consist of a series of thin bridges from

**Fig. 10.** Particulate coating on the cytoplasmic surface of the folded apical membranes of insect rectal papillae. These have also been found in insect rectal pads, Malpighian tubules, insect salivary glands, and insect midgut goblet cells. A particulate material has also been found on the outer surface of the basal infolds of chloride cells of fish (229). ×150,000. Courtesy of Dr. Brij L. Gupta. From *J. Cell Biol.* **29**, 376–382.

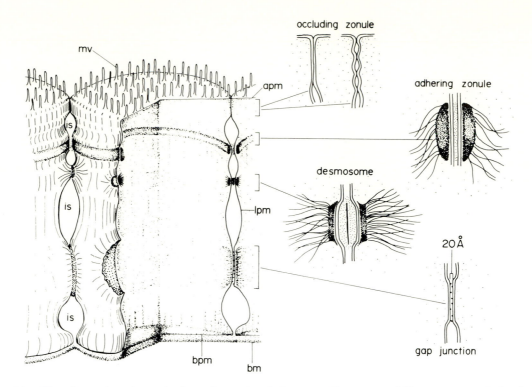

**Fig. 11.** Vertebrate junctions. A three-dimensional diagram illustrating different views of the four major junctional types. Closest to the lumen, the occluding zonule (tight junction) is a region of apparent fusion of the outer leaflets of the plasma membrane to form a seal which completely encircles the cell. There may be a single line of fusion or a series of fused regions separated by spaces. The junction hinders diffusion from the fluid bathing the apical surface into the intercellular space (is). The adhering zonule and desmosome (adhering macula) are sites of adhesion between cells which are not barriers to diffusion because substances can pass through the former or circumvent the latter. Together with the occluding zonule, the adhering zonule and desmosome form the so-called junctional complex which is characteristic of many epithelia. At gap junctions (occluding maculae) there is close apposition (20 Å) of the neighboring cell surfaces but no fusion. The large stippling on the face view of the bisected gap junction illustrates the array of particles which are thought to bridge the 20 Å gap. Gap junctions may function both in adhesion and cell-to-cell communication.

cell to cell (*93, 166, 238, 308*). Septate desmosomes serve to hold cells together and may also provide the low resistance pathways for the passage of ions and molecules from cell to cell.

### Mitochondria

Transporting cells frequently possess an abundance of mitochondria, as expected of cells carrying on oxidative phosphorylation to provide energy for transport. The location of mitochondria within the cells sometimes reflects the location of the transport steps. Thus in some epithelia the mitochondria are located almost exclusively in close contact with one or both cell surfaces (e.g., salivary striated duct, insect goblet cell, maxillary gland tubule, crustacean gill, insect rectum and Malpighian tubule, kidney proximal tubule, stria vascu-

laris). The close juxtaposition of mitochondria to the sites of energy-requiring transport steps may allow a shorter diffusion path for ATP molecules from the mitochondria to the pumps. In other epithelia, mitochondria are found throughout the cytoplasm but are more numerous at one pole of the cell than at the other (e.g., gallbladder, granular cell of toad bladder, rectal gland).

In some epithelia the mitochondria have an unusual morphology that may be related to the high energy demands placed upon them. In *Calliphora* salivary glands the mitochondrial cristae are perforated so that there is a regularly spaced channel system extending through the interior (*197*). This type of mitochondrion has been found in the flight muscle of the same insect (*258*) and in the heart muscle of canaries (*257*), both of which are tissues with a high metabolic demand.

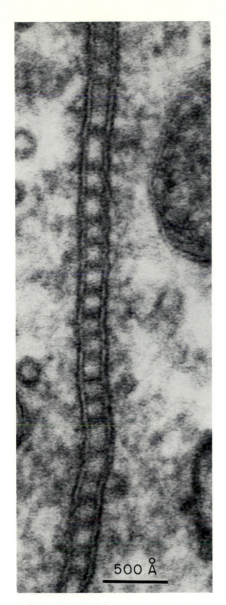

**Fig. 12.** The septate desmosome, consisting of a series of thin lamellae traversing the intercellular cleft between the outer leaflets of opposed plasma membranes. Two different types have been described. ×340,000. Courtesy of Dr. Brij L. Gupta.

## Myoid Elements

Many glands have myoepithelial cells that function in secretion by squeezing acini and regulating the diameter of ducts. In other epithelia, where discrete myoepithelial cells are absent, the transporting cells themselves may be able to contract because of the presence of fine contractile filaments, similar in size

to those of smooth muscle, in the basal regions of the cells. These filaments have been described in kidney tubules (*209, 234*) but may prove to be of widespread occurrence.

## CONCLUSIONS AND PERSPECTIVES

We have attempted to show how the standing gradient model explains the isosmotic fluid transport that occurs in many epithelia, and how it may apply with equal force for both absorptive and secretory epithelia. To place this review in a proper perspective, we must point out that progress in refining the hypothesis will not cease with this writing. Ingenious methods will be devised in an effort to study standing gradients, and much information about the movement of molecules in narrow spaces will be obtained. The model may change significantly, yet even in its present form, the standing gradient hypothesis gives us for the first time a common ground for discussing all of the epithelia described in this volume. The main virtue of the concept is that it allows us to see how structural features of transporting cells fit with their functional properties as measured by physiologists. We follow with great interest the development of alternative proposals (e.g., *43, 90, 228*) even though they have not yet achieved a thorough integration of structure and function.

From this review we have learned that sodium is not the only ion used to generate osmotic gradients that bring about water movement across cell layers. In insects potassium is often used, lactose may be utilized in the mammary gland, and bicarbonate in the pancreas. We find that the structure and function of gland ducts is not always given adequate attention, although ducts may have major roles in the formation of the final secretion. The basal epithelial complex described by Kaye and Lane (*133*) seems to be a universal feature of absorptive epithelia, but its presence has often been ignored in interpretations of physiological data. Finally, the standing gradient hypothesis may also have application to the movement of water in plants (*5*) and may ultimately be applied to the movement of fluids through channel systems within cells, such as the endoplasmic reticulum and Golgi complex.

## OSMOREGULATION AND EXCRETION

Renal and extrarenal organs have dual functions. They excrete nitrogenous and other waste products of metabolism and they maintain the internal fluid or blood at a relatively constant ionic composition and osmolality, even when the external environment changes. This is accomplished by eliminating excesses while conserving those substances that are not abundant in the environment. The complexity of the regulatory task depends on how much the external environment varies from time to time and on how much it differs from the internal environment. Thus the marine habitat is characterized by an abundance of both water and ions, and organs such as salt gland, rectal gland, and fish gills excrete excess ions taken in by drinking seawater. Ions are at very low concentration in fresh water, and organs such as protonephridia, antennal glands, and maxillary glands are involved in ion conservation and elimination of excess water. Anal papillae, crustacean gills, and amphibian epidermis are involved in accumulating ions from dilute surroundings. Water is often a limiting factor in the terrestrial habitat, and the insect Malpighian tubule–rectum system, amphibian urinary bladder, and mammalian and reptilian kidneys are highly efficient in conserving water. The insect goblet cell and aglomerular kidney of marine teleost fish are adaptations that enable these organisms to cope with special problems created by their particular habitats.

## Protonephridia

Protonephridia are the excretory and osmoregulatory organs of certain invertebrates (e.g., Platyhelminthes, rotifers, and nemertines). In freshwater rotifers, the protonephridia function in osmoregulation by conserving solutes and eliminating excess water as a hypotonic urine (24). There are two kinds of protonephridia, flame bulbs and solenocytes. The flame bulb consists of a single cell bearing tufts of cilia. Solenocytes, which

will not be described in detail, have a single long flagellum (23, 186).

Each flame bulb connects to a tubule leading to a contractile bladder. The number of bulbs associated with each protonephridium and the total number of protonephridia varies with different species (215). The flame bulb has a cytoplasmic cap containing the nucleus, and an elongated circular basket. The latter consists of an outer layer of long cytoplasmic projections (cp) connected to each other by strands of glycocalyx, and an inner layer of microvilli (mv). At the end of the basket the cytoplasmic projections are connected to the first cells of the tubule. Cilia extend down the inside of the basket and may project into the tubule lumen. In the tapeworm there are numerous tightly packed cilia (22, 152). Fewer cilia are found in planaria (172) and rotifers (42, 298).

Urine is formed by ultrafiltration across the flame bulb basket followed by reabsorption in the tubule (24). The rate of bladder emptying and the rate of ciliary beating is proportional to the osmotic pressure of the bathing medium (216). Filtration probably depends on the suction created by beating of the cilia (218). The microvilli and cytoplasmic projections may function together as ball and socket valves allowing fluid to enter the basket (a) but resisting flow in the opposite direction (b) when the microvilli are pushed against the outer cytoplasmic projections attached to the basement membrane (bm). The juxtaposition of microvilli and cytoplasmic projections may be maintained by the strands of glycocalyx.

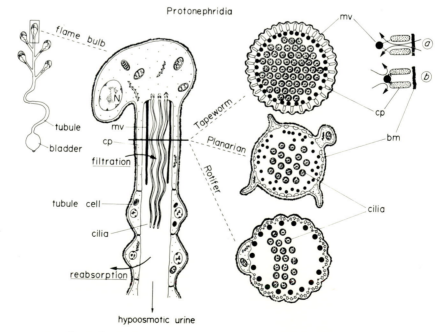

**Fig. 13.** Transverse section through the basket of the protonephridium of a planarian *Dugesia tigrina.* ×20,000. From *Z. Zellforsch. Mikrosk. Anat.* **92,** 509–523. Courtesy of Dr. James A. McKanna.

**Fig. 14.** Protonephridium from the rotifer, *Asplanchna brightwelli.* Transverse section of the flame bulb just below the apical cap. ×28,000. From *J. Ultrastruct. Res.* **29,** 499–524. Courtesy of Dr. Fred D. Warner.

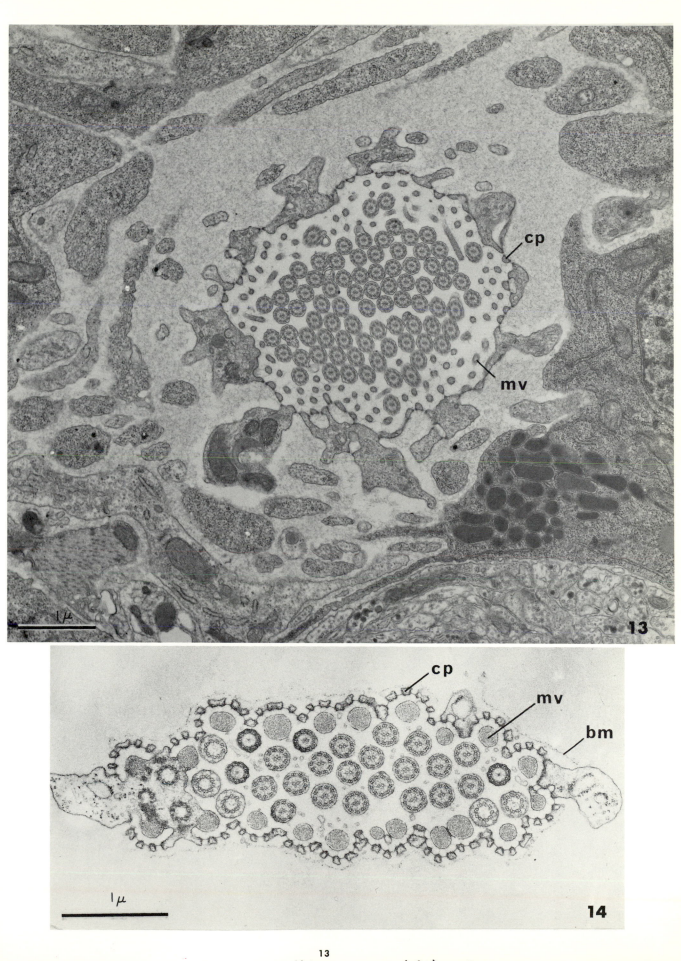

## Invertebrate Segmental Organs

The maxillary and antennal (green) glands of crustaceans and the coxal glands of arachnids are examples of primitive invertebrate segmental excretory organs. In aquatic crustaceans they play a major role in both excretion and osmoregulation (145) whereas in arachnids they are primarily concerned with osmoregulation (159). The ultrastructural organization of segmental organs in crustaceans (153, 241, 285, 286), arachnids (99, 109, 225), myriapods (105) and insects (2, 106) is similar, and a general description is provided here.

Segmental organs consist of two main parts, a proximal closed end sac, and a labyrinth made up of a single coiled tubule. In some cases the tubule is connected to a bladder. The end sac resembles Bowman's capsule of the vertebrate kidney (p. 34), as the cells have many interdigitating foot processes (fp) resting on the basement membrane (bm). Between adjacent foot processes are narrow filtration slits that are bridged by thin septa (Fig. 16). The tubule cells have extensive membrane elaboration on both surfaces (Fig. 17). The apical membrane has numerous microvilli (mv) whereas the basal and lateral membranes have many deep infoldings closely associated with mitochondria. Septate desmosomes (sd) join the cells in the apical region.

Fluid formation depends on ultrafiltration at the end sac followed by reabsorption and/or secretion of ions and nonelectrolytes in the tubule. During ultrafiltration fluid passes through the basement membrane, between the foot processes, and finally percolates through the intercellular spaces into the lumen (arrows). Filtration in Crustacea probably depends on the hydrostatic pressure of the blood, while in the coxal glands of arachnids filtration seems to result from contraction of muscles attached to the end sac (159).

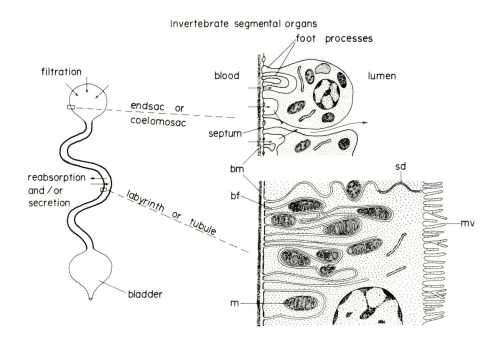

Invertebrate segmental organs

Fig. 15.  The end sac of the maxillary gland of *Artemia salina*. ×10,000. Courtesy of Dr. Greta E. Tyson.

Fig. 16.  High magnification of the filtration slits of the end sac. ×43,000. Courtesy of Dr. Greta E. Tyson.

Fig. 17.  Labyrinth cells in the antennal gland of the fiddler crab *Uca mordax*. ×18,875. From *J. Morphol.* **125**, 473–495. Courtesy of Dr. Lowell E. Davis.

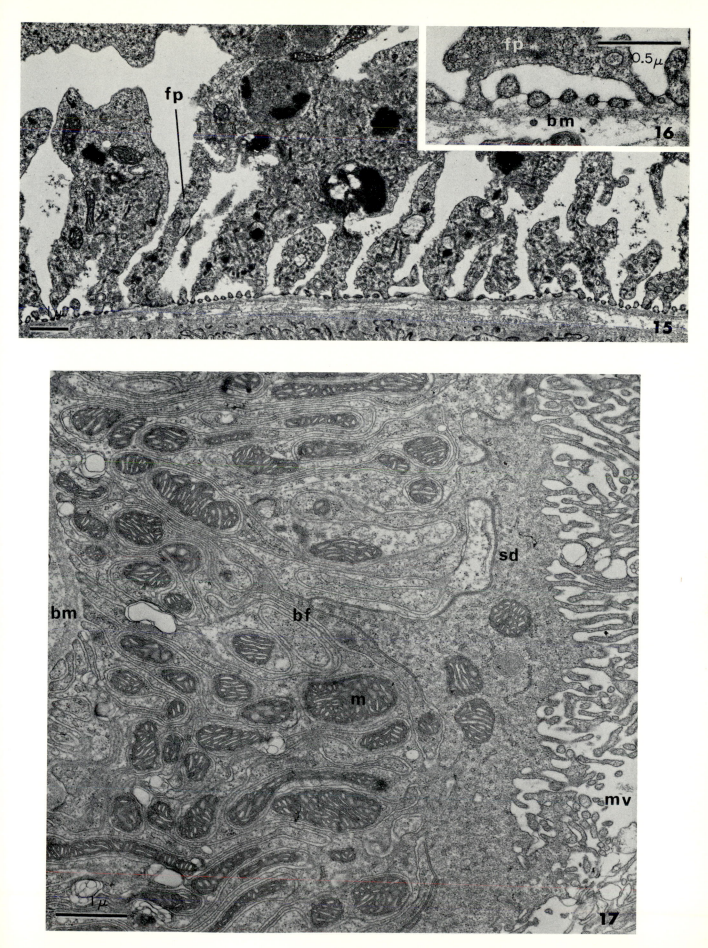

### Invertebrate Extrarenal Organs

Osmoregulation in freshwater invertebrates involves uptake of salts across the body surface to replace loss to the dilute environment. The salt-absorbing organs most frequently studied are the anal papillae of mosquito larvae and the gills of certain Crustacea. In both, the absorptive surfaces are flattened sacs (lamellae) lined with a single layer of epithelial cells. In gills the lamellae are occasionally traversed by pillar cells, and the blood space communicates with the afferent branchial vessels (af. b.v.) and the efferent branchial vessels (ef. b.v.). The absorptive cells (stippling) are concentrated around the afferent blood vessels (46) as are the chloride cells in fish gills (p. 32).

The absorptive cells of crustacean gills (Fig. 19) and mosquito anal papillae (Fig. 18) have a similar ultrastructural organization (45, 46). The apical plasma membrane is highly folded (af) and protected by cuti-

cle (c). Mitochondria (m) lie adjacent to the folded region but do not penetrate between the folds. The basal plasma membrane is also folded (bf) with large mitochondria adjacent to the membranes. A thin basement membrane (bm) separates the cells from the hemolymph (heme). Tracheoles (t) ramify among the basal infolds of the anal papillae.

Sodium and chloride are actively absorbed by mechanisms which can be independent but which are more efficient when both ions are present. In the former case, electroneutrality is probably achieved by exchanging external sodium for hydrogen or ammonium ions, and external chloride for hydroxyl or bicarbonate ions (167, 218, 265). The absorption of sodium seems to involve active ion pumps working in conjunction with exchange-diffusion mechanisms. The latter may be located on the apical membrane while the active pumps are probably situated on the infolded basal membrane (279).

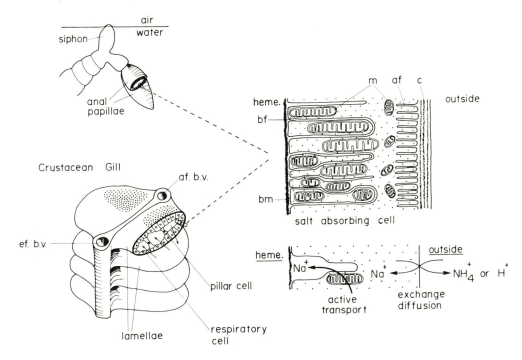

Invertebrate extrarenal organs

Mosquito Anal Papillae

Crustacean Gill

salt absorbing cell

**Fig. 18.** Anal papillae of the larval mosquito *Culex quinquefasciatus*. ×24,000. From *J. Cell Biol.* **23**, 253–263. Courtesy of Dr. D. Eugene Copeland.

**Fig. 19.** The gill epithelium of the blue crab *Callinectes sapidus*. ×8000. Courtesy of Dr. D. Eugene Copeland.

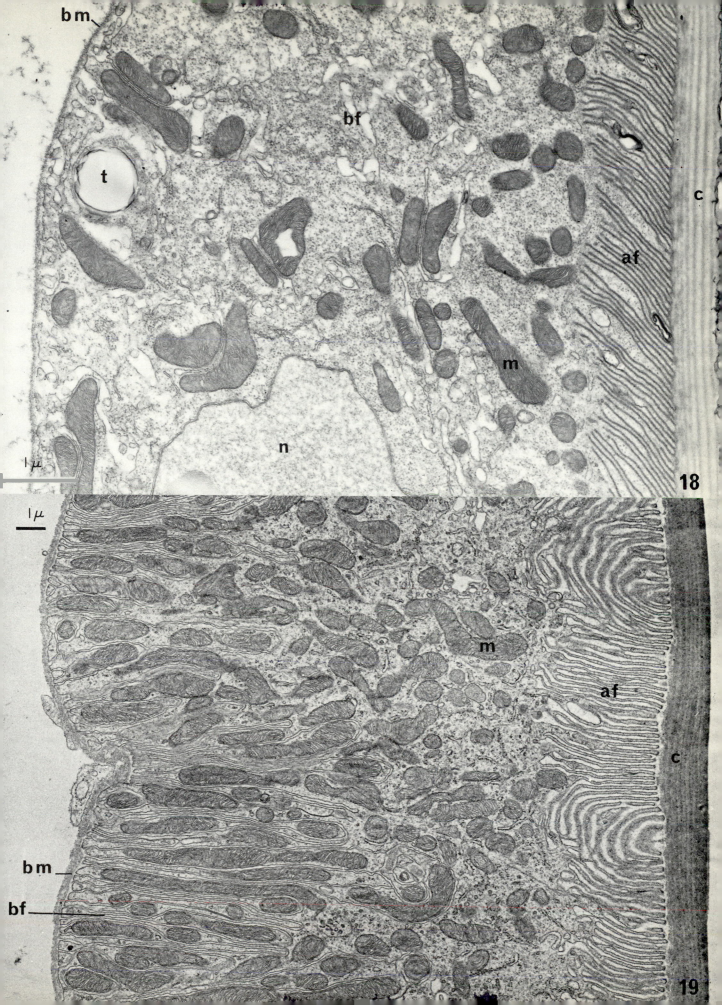

## Insect Malpighian Tubule

In insects osmoregulation and excretion are two-step processes involving the secretion of a primary urine by the Malpighian tubules and reabsorption of required substances by the rectum. The tubules ramify in the hemocoel or body cavity and drain into the gut at the junction between midgut and hindgut. The number and form of the tubules varies widely among the different insects. In some the tubules are of uniform structure along their length. For example, in *Calliphora* the tubules consist of two cell types that intermingle along the whole length (*19*). In *Rhodnius* the tubules are differentiated into upper (us) and lower (ls) segments (*303*). In certain insects the tubules consist of several different regions and the distal ends insert on the rectum to form a cryptonephric arrangement (*98, 224, 259*).

The secretory cells of Malpighian tubules (e.g., the primary cells of *Calliphora* and the upper segment cells of *Rhodnius*) are similar in all insects examined. Both basal and apical surfaces are highly folded with mitochondria (m) frequently associated with both surfaces (Fig. 20). In some tubules the mitochondria extend into the microvilli (Fig. 22, mv) (*10, 19*). The cells are joined by septate desmosomes (sd) and often contain large concretions and clear vacuoles (v).

Insect hemolymph (heme) is not under sufficient hydrostatic pressure for ordinary ultrafiltration. Urine formation is probably accomplished by local osmotic gradients set up in the long narrow basal channels and also the channels between microvilli (*19*). Osmotic gradients (heavy stippling) created by ion uptake (solid arrows) from the basal channels may draw water (broken arrows) through the basement membrane (bm) and cell by osmotic filtration (see Fig. 5). Large protein molecules would be excluded by the basement membrane (*198*). The primary urine in most insects contains much more potassium than sodium and it is thought that active potassium transport is the prime mover for fluid secretion (*221, 223*). Smaller molecules (amino acids, sugars, wastes) are carried with the fluid flow while organic acids (e.g., phenol red) are actively secreted (*16*).

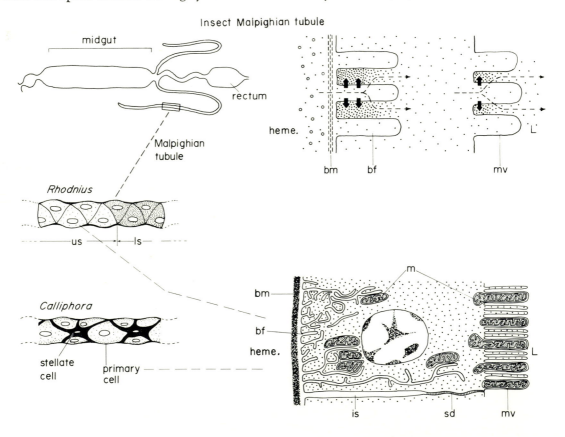

Insect Malpighian tubule

**Fig. 20.** The Malpighian tubule of the cockroach *Periplaneta americana.* ×12,500.

**Fig. 21.** Basal interdigitations of the Malpighian tubule of the cockroach. ×30,000.

**Fig. 22.** Microvilli containing mitochondria at the apical surface of the Malpighian tubule of *Calpodes ethlius.* ×30,000. Courtesy of James T. McMahon.

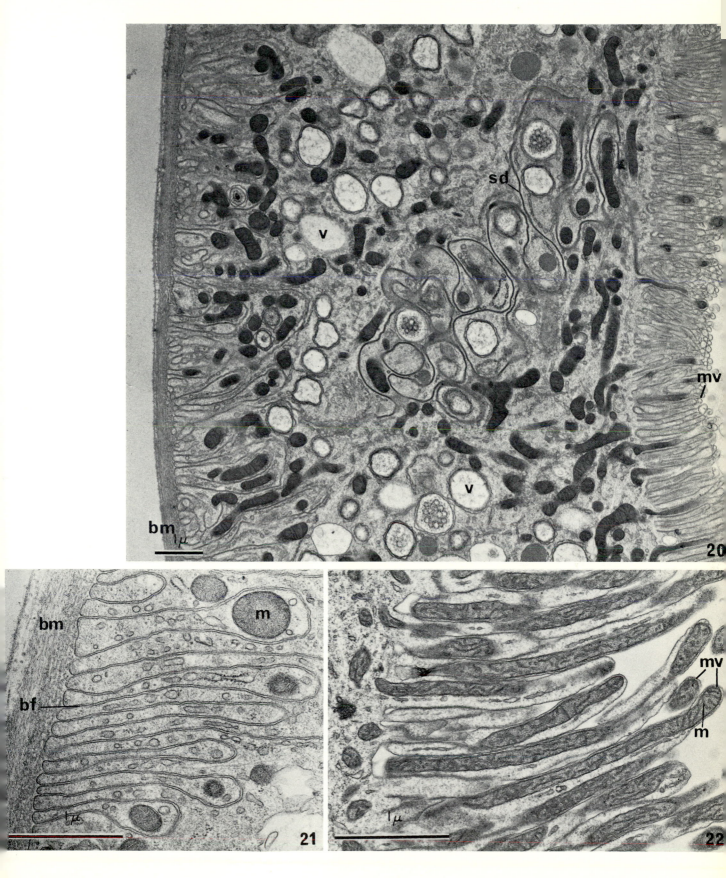

## Insect Rectum

The rectum of insects functions in regulation by selectively reabsorbing water, ions, and probably amino acids and sugars that are secreted by the Malpighian tubules (222, 301). Waste products and excess ions are left behind and excreted. The structural complexity of the rectal epithelium varies among the insects and seems to be related to the extent to which water can be absorbed to concentrate the feces. In many insects absorption is accomplished by specializations of the rectal epithelium known as rectal pads (Dermaptera, Orthoptera, Carabidae) or papillae (Siphonaptera, Diptera, adult Lepidoptera). These are structurally the most complex of the transporting epithelia yet studied.

The absorptive cells of rectal papillae have folded apical membranes (Fig. 26, af) but the most extensive membrane elaboration is found on the lateral membranes, which are folded into a series of stacks (Fig. 25, st) of parallel membranes closely associated with mitochondria (17, 101). ATPase activity is closely associated with these stacks (18). In rectal pads the apical membrane (Fig. 24) is more elaborate and mitochondria are associated with both apical and lateral (Fig. 23) surfaces (199). In both epithelia septate des-

mosomes (sd) are present near both apical and basal surfaces, so that the extensive intercellular spaces (is) communicate with the hemolymph at only a few regions. In rectal papillae, fluid leaving the intercellular spaces (is) enters the infundibulum at the apex before flowing toward the base, where it passes into the hemolymph through valves. Fluid absorbed by rectal pads drains via indentations of the basal surface where tracheae (t) penetrate, through a sinus, and into the hemolymph via gaps in the muscle layer.

The cuticle (Figs. 24 and 26, c) lining the rectum functions as a molecular sieve, allowing only small molecules access to the apical surface (211). Active transport of ions into the intercellular spaces (a) elevates the osmotic concentration there (295–297) and brings about a flow of water from the lumen (b). Ordinarily ions pumped into the spaces are replaced by ions entering from the lumen (c). However, some insects are able to absorb water when no ions are present in the lumen (210). The ions may be obtained from the infundibulum or sinus (d); or the papillae may recruit them from the hemolymph (e) (see also Fig. 7). In these cases water uptake (b) continues although the ions used to generate the osmotic gradients are not obtained from the lumen.

Insect rectum

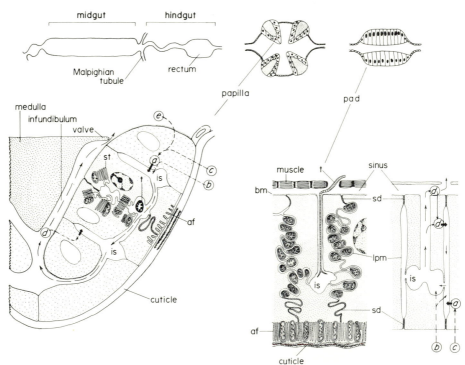

**Fig. 23.** Lateral membrane foldings associated with mitochondria in the rectal pads of the cockroach *Periplaneta americana*. ×35,000.

**Fig. 24.** Apical surface of a rectal pad of the cockroach. ×5300. From (199).

**Fig. 25.** Lateral membrane stacks found in the rectal papillae of the blowfly *Calliphora erythrocephala*. ×70,000. Courtesy of Dr. Brij L. Gupta.

**Fig. 26.** Folded apical surface of blowfly rectal papilla. ×24,000. From *J. Cell Sci.* **2**, 89–112. Courtesy of Dr. Brij L. Gupta.

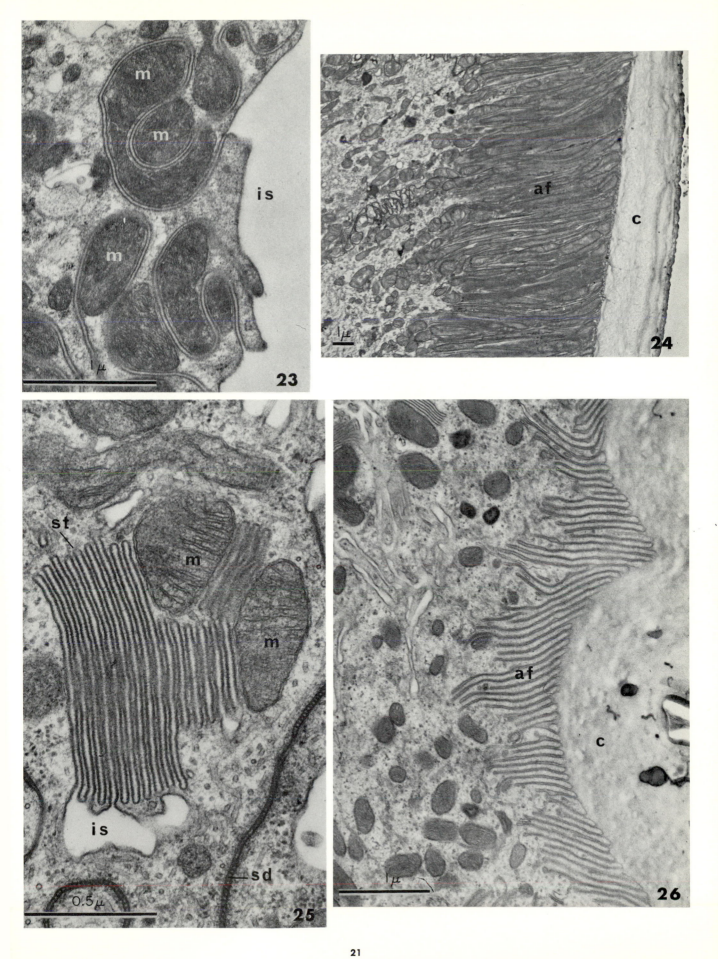

## Insect Goblet Cell

In addition to normal columnar cells (p. 58) the intestine of certain insects (larval Lepidoptera and Trichoptera) also have goblet cells. These cells have only a superficial resemblance to vertebrate goblet cells (Fig. 77). Herbivorous insects obtain large amounts of potassium in their diets and goblet cells may function in ionic regulation by secreting potassium from the blood into the intestinal lumen (103).

Goblet cells have a large cavity (gc) formed by an invagination of the apical plasma membrane (4, 260). The inner surface of the cavity is lined with long thick microvilli, each containing a single long mitochondrion (Figs. 28, 29). The cytoplasmic surface of the plasma membrane of the microvilli is lined with particles (p. 8). The opening of the goblet cavity is guarded by villous-like projections of the apical membrane (Fig. 27). Goblet cells connect to columnar cells by gap junctions (occluding macula, om) and septate

desmosomes (Fig. 30, sd). Toward the basal region, the lateral membrane lacks junctions and is arranged in flattened parallel folds. The basal plasma membrane has some infoldings (bf) which are unusual in that they tend to lie parallel to the basement membrane (bm) forming a compartment with relatively few openings into the hemolymph.

The mechanism of potassium secretion has been studied using isolated midgut preparations which maintain a potential (lumen positive) for many hours (102–104, 307). Most of the short-circuit current is carried by a potassium transport system that is insensitive to ouabain and completely independent of other cations. The potential may thus be generated by an electrogenic potassium pump. Potassium may enter passively across the basal surface (a) to be extruded into the goblet cavity by the electrogenic pump (b) (306). Energy for this process would be derived from the mitochondria which are intimately associated with this apical surface.

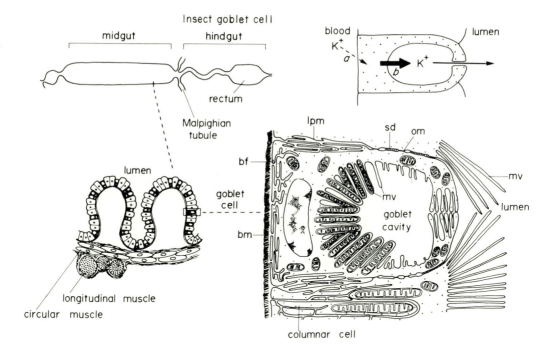

Fig. 27.  Apical surface of the goblet cell of *Antheraea pernyi*, a lepidopteran silk moth. ×26,000. Courtesy of Dr. Brij L. Gupta.

Fig. 28.  Basal region of the goblet cell. ×13,000. Courtesy of Dr. Brij L. Gupta.

Fig. 29.  The entire goblet cell of the larval silk moth. ×4200. Courtesy of Dr. Brij L. Gupta.

Fig. 30.  The junction between goblet cell and columnar cell. ×157,000. Courtesy of Dr. Brij L. Gupta.

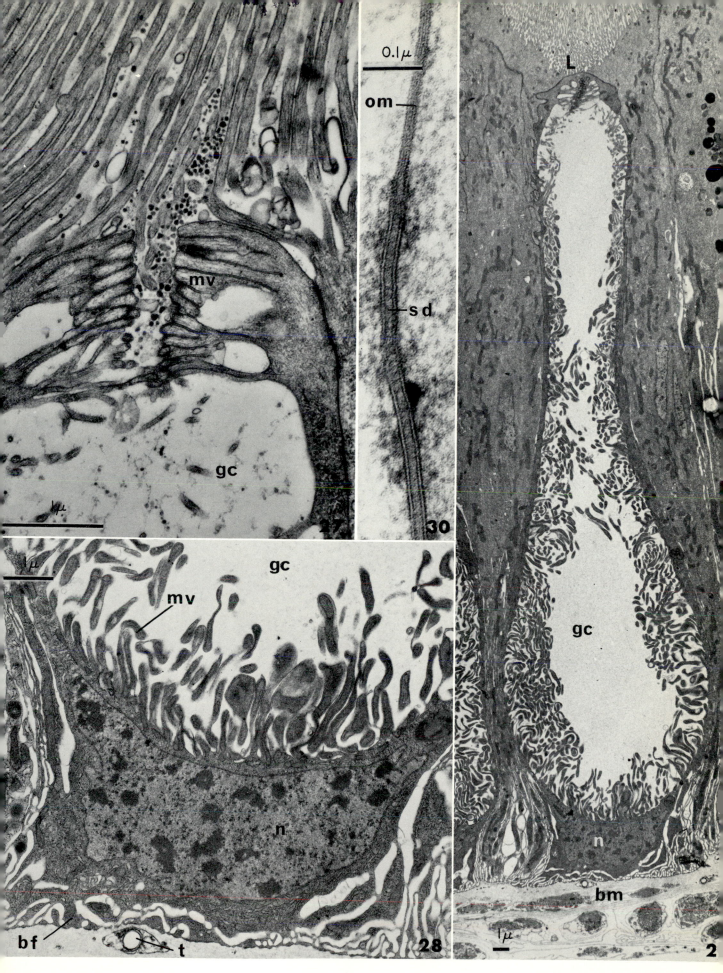

## Amphibian Epidermis

Frog skin has been used extensively as a model system for studying ion transport. Under normal conditions the skin functions in osmoregulation by absorbing sodium and chloride from the dilute external environment.

The structure of the skin is complicated by the presence of four regions, the stratum corneum, stratum granulosum, stratum spinosum, and stratum germinativum, composed of different cell types (84–86, 292, 293). The layers represent stages in the transition from the germinative layer facing the basement membrane to the fully keratinized cells of the stratum corneum facing the outside. The cells are held together by desmosomes (d) which increase in number towards the outermost layers. The layers closest to the outside (stratum corneum and stratum granulosum) have occluding zonules (oz) but these are absent from the other two layers. However, these inner layers are linked to each other and to the stratum granulosum by gap junctions (occluding maculae, om). The intercellular space (is) surrounding these three cell types is thus continuous. The intercellular space between cells in the stratum granulosum is narrow but widens around the stratum spinosum and stratum germinativum, where it is often highly tortuous due to extensive interdigitations between cells.

Active sodium uptake involves passive influx across an osmotic barrier followed by active transport into the blood space (147). The location of the osmotic barrier and the transport step is uncertain owing to the structural complexity of the system. Sodium probably diffuses in passively across the stratum corneum and the outer surface of the stratum granulosum (a) (294) and is then transported out into the intercellular spaces surrounding the stratum granulosum (b). Sodium may also diffuse from cell to cell across the gap junction (om) into the stratum spinosum and stratum germinativum (c) which may also extrude sodium into the intercellular spaces. Sodium finally diffuses out into the blood through the narrow spaces between adjacent cells of the stratum germinativum (d).

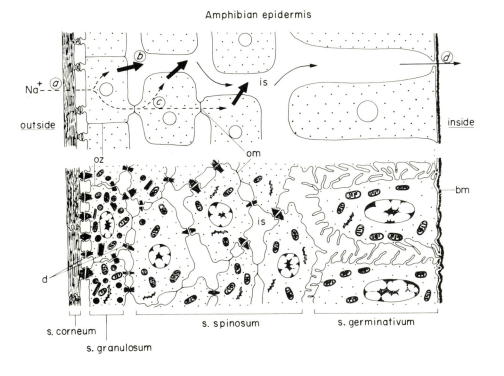

Amphibian epidermis

Fig. 31. Survey picture of the whole frog skin epithelium. Stratum corneum (SC) at left. ×4100.

Fig. 32. Desmosome between the stratum corneum and s. granulosum (SG). ×80,000.

Fig. 33. Occluding zonule between adjacent cells of the stratum granulosum. ×80,000.

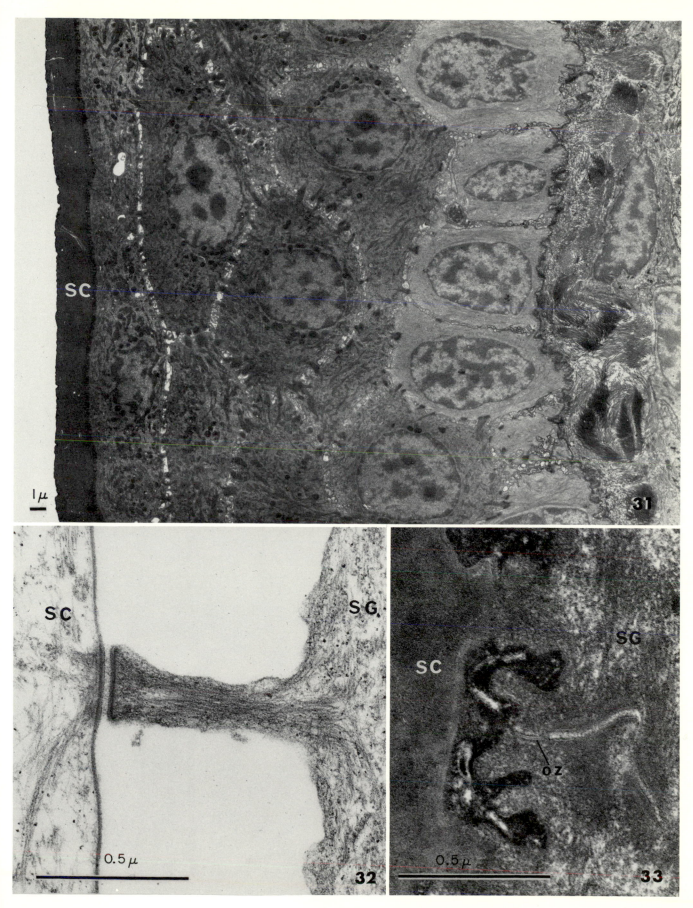

## Amphibian Urinary Bladder

The urinary bladder of amphibia plays an important role in osmoregulation by functioning as a storage organ for water (12). Neurohypophysial hormones regulate water absorption from the hypoosmotic urine stored in the bladder. Sodium and urea can also be absorbed from the bladder lumen.

The bladder is composed of four cell types: granular, mitochondria-rich, mucous, and basal. Granular cells (gc) predominate facing the lumen, with only occasional mitochondria-rich, or mucous cells (41, 61, 88, 207). The flattened basal cells (bc) lie adjacent to the basement membrane (bm) and have no contact with the lumen. Very small projections of the granular cells pass between the layer of basal cells to touch the basement membrane. The extensive intercellular space (is) formed between these two cell types becomes a narrow channel on the side near the blood making a separate compartment within the epithelium. Most of the serosal surface of the granular cells faces this enclosed intercellular space rather than the blood. The complex structural relationship between granular and basal cells has not usually been considered in interpretation of hormonal and physiological studies.

The apical plasma membrane is the main permeability barrier between lumen and blood. ADH (antidiuretic hormone), acting via cyclic AMP, increases the permeability of this membrane to both ions and water (12, 142). Water enters the cell from the hypoosmotic urine (a) and moves into the blood, probably via the intercellular spaces (b). Sodium entering passively from the lumen is extruded from the cell by active sodium pumps located on the lateral membranes of the granular cells (c). Most substances must be routed through the intercellular space when in transit from lumen to blood. It has been reported that the space distends during active absorption (34, 35, 97, 201, 204).

The toad bladder is ontogenetically and functionally analogous to the collecting ducts of kidney (p. 38) where an enclosed intercellular compartment is also involved in fluid absorption.

Amphibian urinary bladder

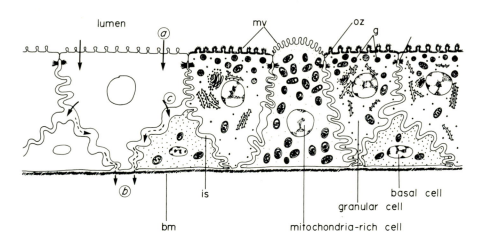

Fig. 34. Granular and basal cells of toad bladder. ×21,000. Courtesy of Mr. Peter Heap and Dr. Douglas R. Ferguson.

Fig. 35. Mitochondria-rich cell of the toad bladder. ×11,000. Courtesy of Dr. Donald R. DiBona.

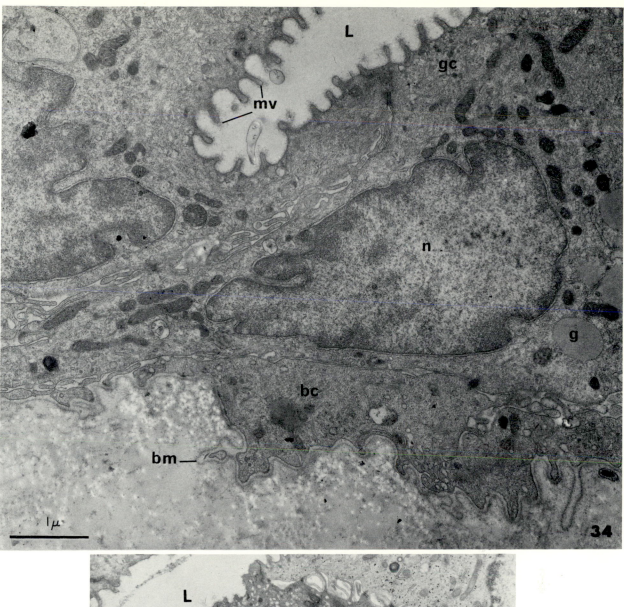

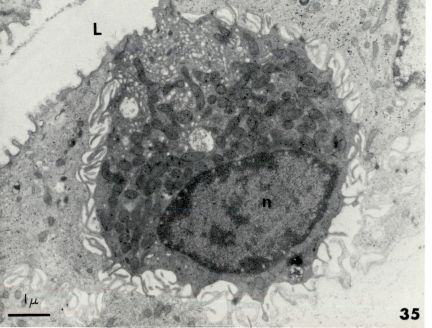

## Salt Gland

The salt gland of marine reptiles and birds secretes the excess salt obtained from food and water (243, 244). The secretion is a hyperosmotic sodium chloride solution (ranging from 0.5–1.0 $M$ in various species) with trace amounts of other ions and nonelectrolytes. The gland is under nervous (parasympathetic) control and is active only after the animal ingests a salt load (245). Secretory activity can be regulated experimentally by administering acetylcholine or by electrical stimulation of the secretory nerve which innervates the tubule cells (82).

Although the salt gland of birds is derived from nasal glands while reptiles have modified lacrimal glands, their gross morphology is similar (1, 81). The gland consists of many long lobes each with a central lumen from which many closely packed branching tubules radiate. The tubules are invested with blood vessels arranged so that blood flows (arrows) in a direction opposite to the tubular secretion (dashed arrows), establishing a countercurrent flow system. The narrow tubule lumen is surrounded by pyramidal cells (Fig. 36). In reptiles, membrane elaboration is achieved through extensive infolding and interdigitation of neighboring intercellular membranes to form a tortuous intercellular channel (is) (1, 161). In birds, however, both lateral and basal membranes are infolded to produce long cytoplasmic compartments containing mitochondria (Fig. 37) (69, 78, 148). In both, the lumen (L) is narrow and the apical surface is only slightly amplified by a few short microvilli. There is a junctional complex (jc) near the apical surface (Fig. 38).

The hyperosmotic fluid is secreted across the apical plasma membrane, since the intracellular sodium concentration is low (208). It has been suggested (208) that when the cell is quiescent, sodium is returned to the blood by a sodium/potassium exchange pump on the basal surface. Acetylcholine (ACh) may stimulate the gland by increasing the passive entry of sodium and the subsequent increase in intracellular sodium concentration may activate an electrogenic pump on the apical plasma membrane to secrete sodium into the lumen. The extensive infolding of the basal and lateral membranes probably facilitates uptake of ions and metabolites from the extracellular fluid, while the small apical surface area minimizes the osmotic flow of water as ions are extruded into the lumen.

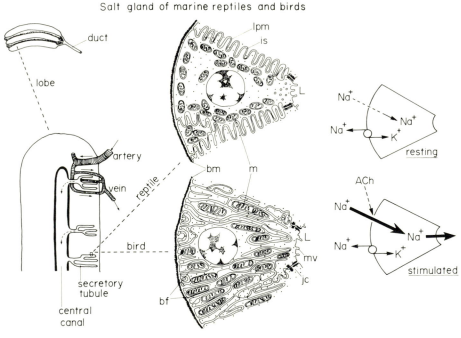

Salt gland of marine reptiles and birds

Fig. 36. A salt-secreting cell from a duck. Lanthanum treated to show the folding of lateral and basal membranes. ×10,250. Courtesy of Dr. Stephen A. Ernst.

Fig. 37. The basal region of a specimen in which the spaces of the basal labyrinth are distended. ×11,500. Courtesy of Dr. Stephen A. Ernst.

Fig. 38. High magnification of the narrow lumen that is bounded by six cells joined by desmosomes and occluding zonules. ×23,800. From *J. Cell Biol.* 40, 305–321. Courtesy of Dr. Stephen A. Ernst.

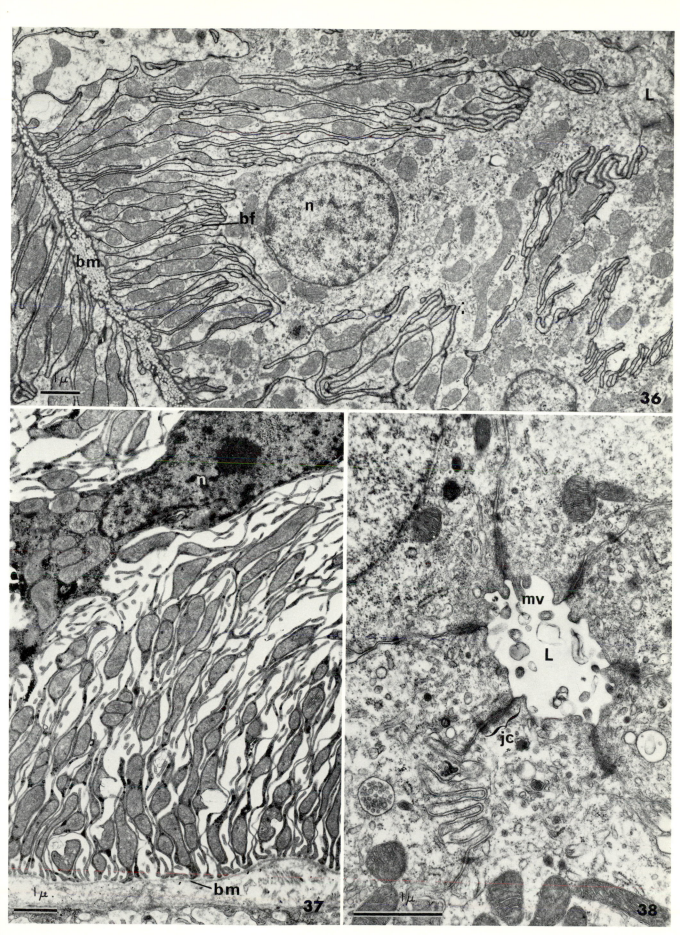

29

## Rectal Gland of Elasmobranch Fish

Rectal glands in elasmobranch fish function in osmo-regulation by secreting hyperosmotic sodium chloride into the rectum (*31*). The glandular region has an outer capsule of connective tissue and smooth muscle surrounding an inner parenchyma of branched secretory tubules (*39, 48*). Hyperosmotic sodium chloride passes from the secretory tubules into the rectal gland lumen which is lined by a stratified epithelium several cells deep. Blood from the rectal gland artery circulates around the tubules and accumulates in large sinuses lying near the lumen, and finally flows into the rectal gland vein. Unlike the salt gland, there is no counter-current system because the blood flow is in the same direction as the secreted fluid. The rectal gland is not innervated.

Each branched secretory tubule has a narrow lumen surrounded by a single cell type (Fig. 39) (*27, 70,*

*149*). Most of the basal plasma membrane is straight but there are occasional convoluted basal infoldings (bf). Lateral membranes are also highly folded as adjacent cells interdigitate (Fig. 40). In the apical region there are numerous adhering zonules (az) but no occluding zonules (*160*). The apical membrane (apm) has a few short microvilli (mv). This surface is unusual in that there are numerous interdigitations with apical extensions of neighboring cells (heavy stippling). Large numbers of mitochondria (m) are packed in the cyto-plasmic compartments between basal infolds (bf) and lateral membranes (lpm).

The mechanism of secretion may resemble that in the salt gland just described. The stratified epithelium lining the central lumen may prevent the hyperosmotic fluid from reequilibrating with blood in the large sinuses and may also modify the primary tubular secretion (*28*).

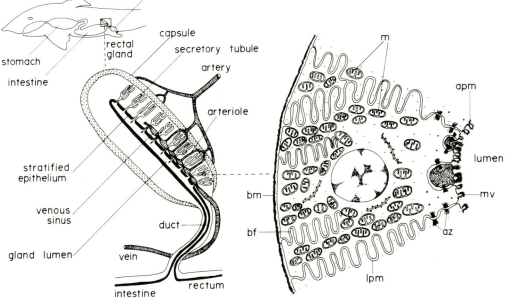

**Fig. 39.** Secretory cell from the rectal gland of the spiny dogfish *Squalus acanthias*. ×8500. Courtesy of Dr. Ruth Ellen Bulger.

**Fig. 40.** Highly folded lateral membranes of the rectal gland of an elasmobranch fish. ×22,000. From *Z. Zellforsch. Mikrosk. Anat.* **74**, 123–144. Courtesy of Dr. H. Komnick.

**Fig. 41.** Apical surface of the rectal gland of the spiny dogfish. ×19,000. Courtesy of Dr. Ruth Ellen Bulger.

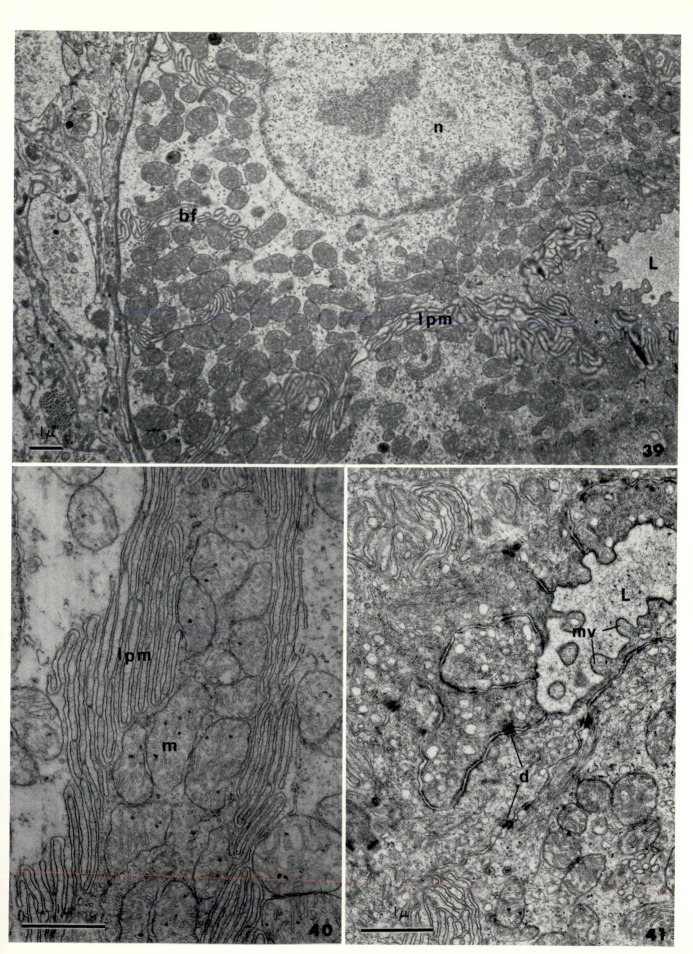

## Fish Gills

The high permeability of the respiratory surfaces of aquatic animals imposes a continuous osmotic stress. Freshwater fish counteract the osmotic influx of water by excreting a dilute urine and absorbing salts through the gills. In marine fish, water loss to the hypoosmotic environment is counteracted by drinking sea water and eliminating the extra solute through the gills (175).

Gill branches are made up of branchial arches from which a series of gill filaments radiate. The main respiratory exchange surface is located on the leaf-like secondary lamellae containing abundant afferent (A) and efferent (E) blood vessels. The epithelium lining the secondary lamellae is composed of one to two layers of flattened cells. Cells of the outer layer are connected by occluding zonules (oz) and desmosomes (d) near the surface, and the lateral plasma membranes interdigitate (117, 187). Cells of the inner layer are less specialized and resemble the basal cells in amphibian bladder (p. 26).

The surface of the gill filament is composed of flattened epithelial cells (Fig. 43) similar to those of the secondary lamellae, interspersed with bulbous mucous cells and chloride cells (Fig. 44). The latter (shaded) are clustered around the afferent blood vessel (A) (44). Chloride cells are characterized by an extensive tubular system (Fig. 42) which ramifies throughout the cell and is closely associated with numerous mitochondria (140, 212, 213, 229, 266). The tubular system is open to the intercellular space (arrows). Numerous vesicles are clustered around the apical membrane. The latter is invaginated to form a cavity (*, Fig. 44) in fish adapted to seawater. The cavity is absent in freshwater-adapted fish (213, 272).

Keys and Willmer (143) proposed that the chloride cells secrete salts in fish adapted to seawater. Gills of fish adapted to fresh water absorb solutes (150) although it is not known if this is accomplished by the chloride cells or by the epithelial cells covering the secondary lamellae (142).

Fish gills

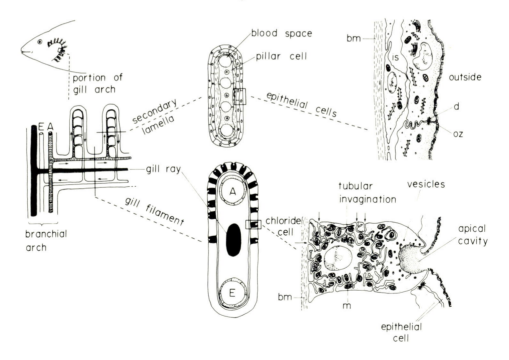

Fig. 42.  Basal surface of the chloride cell. ×34,000. Courtesy of Dr. Charles W. Philpott.

Fig. 43.  The epithelial cell of a gill filament. ×11,250. Courtesy of Dr. Charles W. Philpott.

Fig. 44.  The chloride cell of *Fundulus*. ×4500. Courtesy of Dr. Charles W. Philpott.

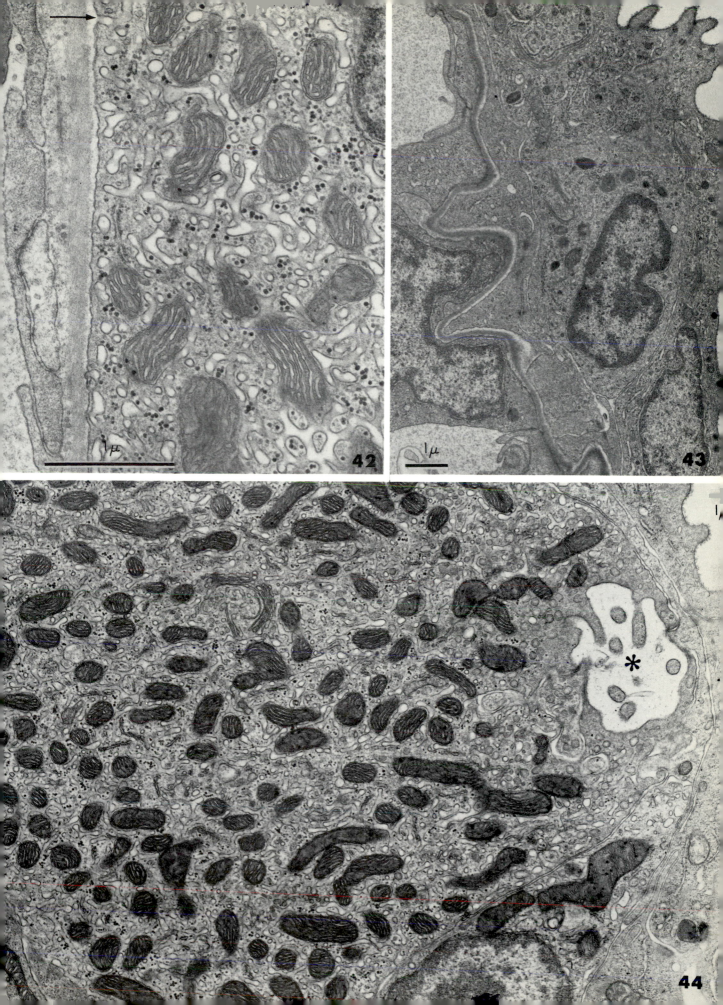

## Mammalian Kidney

The terrestrial habitat presents the most severe osmoregulatory problems. Birds and mammals have kidneys that can produce a concentrated or dilute urine, depending on conditions. The basic functional unit is the nephron, consisting of a glomerulus plus a long, thin, convoluted tubule. A single kidney may contain millions of nephrons. In mammals about one-fourth of the blood pumped by the heart goes to the kidneys where most of it flows through thin capillaries of the glomeruli. The basic mechanism of urine formation is glomerular ultrafiltration followed by tubular reabsorption. When the human kidney is conserving water (dehydrated or antidiuretic condition) about 125 ml of ultrafiltrate is formed each minute, but only 1 ml of urine enters the bladder. The remaining 124 ml are reabsorbed as the fluid flows through the various segments of the nephron and collecting ducts (214).

### Glomerulus

The capillaries (cap) of the glomerulus share a common basement membrane (bm) with the glomerular epithelium (Fig. 47), which is composed of irregularly shaped cells with interdigitating foot processes (fp). Between adjacent foot processes are narrow filtration slits about 250 Å wide (87, 157, 181). Hydrostatic pressure of the blood forces fluid through the capillary wall, basement membrane, and filtration slits, into the urinary space (us) within Bowman's capsule. The resulting ultrafiltrate has the same composition as plasma, except that it is free of the larger proteins. The combination of basement membrane and filtration slits probably acts as a molecular sieve. The ultrafiltrate from the urinary space flows directly into the proximal tubule.

### Proximal Tubule

The proximal tubule reabsorbs much of the water, salts, and small protein molecules filtered by the glomerulus. Organic acids and bases are secreted from the blood into the lumen. The cells have elaborate basal interdigitations and apical microvilli (mv) (158, 227, 273, 274) as well as a highly developed lysosomal apparatus (ly) (77, 179) involved in the intracellular degradation of absorbed proteins. Isosmotic water absorption is probably accomplished by active transport of sodium into the basal labyrinth with water moving passively down the osmotic gradient and chloride diffusing down its electrochemical gradient. The pump is electrically neutral as in the gallbladder (p. 52). Salt and water uptake across the microvillate border concentrates proteins at the bases of the microvilli where they are taken in by pinocytosis. The occluding zonules (oz) are exceptionally short (83) and may represent the site of passive ion permeation across the epithelium (see p. 8).

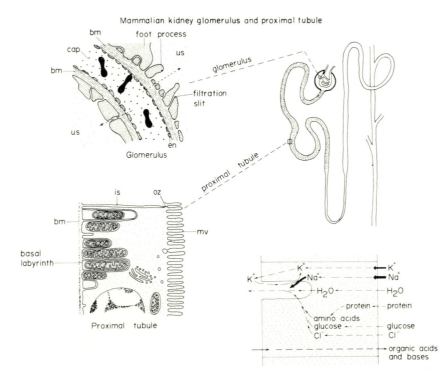

Mammalian kidney glomerulus and proximal tubule

**Fig. 45.** The proximal tubule of the kidney of a rat. ×12,700. Courtesy of Dr. Ruth Ellen Bulger.

**Fig. 46.** Basal labyrinth of the rat kidney proximal tubule. ×21,250. Courtesy of Dr. Ruth Ellen Bulger.

**Fig. 47.** The filtration apparatus of the glomerulus of the rat kidney. ×54,900. Courtesy of Dr. Ruth Ellen Bolger.

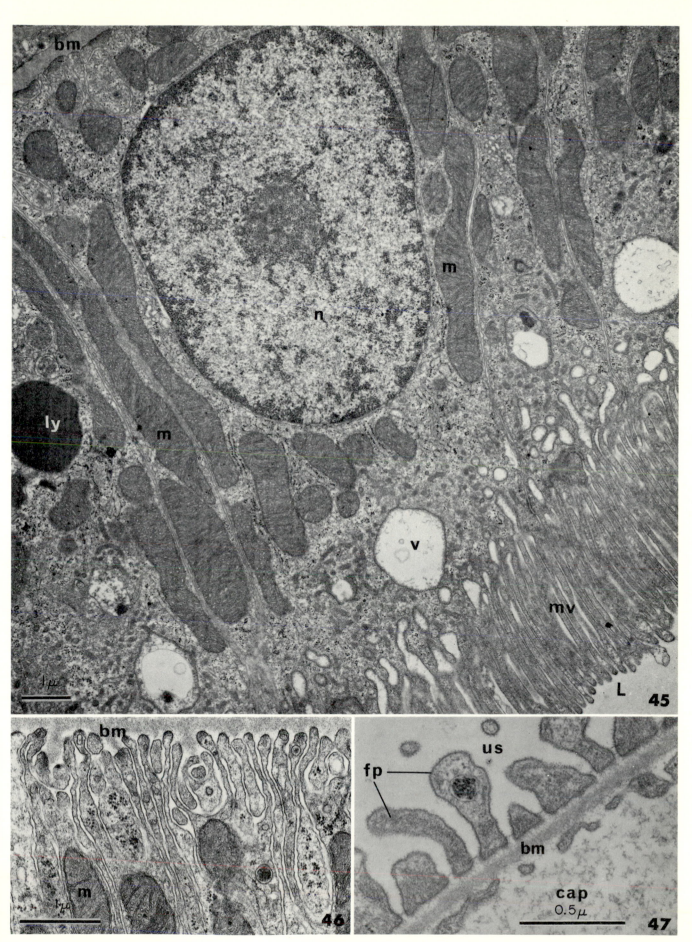

## Distal Tubule

Sodium reabsorption from the distal tubule renders the urine progressively less concentrated. The distal tubule is less permeable to water than the proximal tubule, and sodium reabsorbed is not accompanied by a proportional flow of water (hyperosmotic reabsorption). Chloride reabsorption is mainly passive, while potassium and hydrogen enter the urine, the latter rendering it acidic. Ammonia produced in the cells by deamination of glutamine and other amino acids diffuses into the blood and into the urine. However, ammonia entering the urine is "trapped" there because it is rapidly converted to the relatively nondiffusible ammonium ion. The distal tubule (Fig. 48) has fewer microvilli and a less elaborate basal labyrinth than the proximal tubule, consistent with its role in hyperosmotic absorption (*158, 227, 274*). Also it has longer occluding zonules (oz) than the proximal tubule, providing a possible basis for its lower ion permeability.

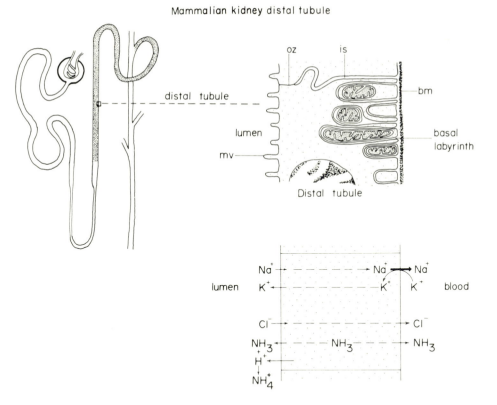

Mammalian kidney distal tubule

**Fig. 48.** The distal tubule of a rat kidney. ×13,450. Courtesy of Dr. Ruth Ellen Bulger.

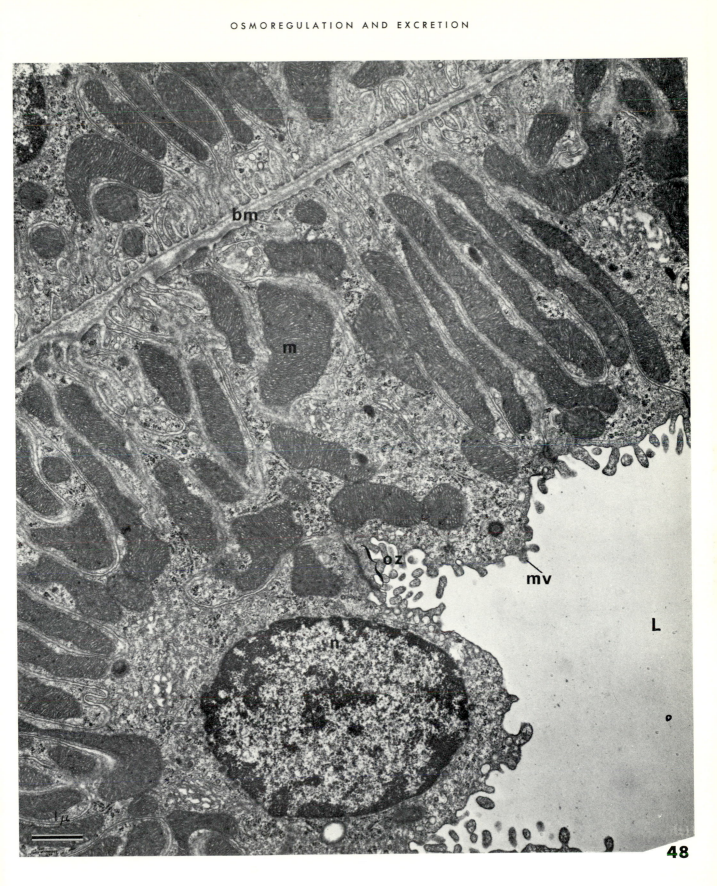

## Loop of Henle

The loop of Henle functions to create a linear osmotic gradient in the interstitial space by countercurrent multiplication. Both ascending and descending limbs are lined by simple squamous cells with few organelles (*127, 200, 227*). Ascending limb cells are joined by ordinary occluding zonules (Fig. 50, oz). Descending limb cells in rabbit kidney have large spaces up to 70 Å wide between them (*52*, Fig. 49 arrows). This has not been confirmed for rat kidney (*30*). The descending limb is permeable to sodium and water while the ascending limb is not. Sodium pumped out of the ascending limb increases the concentration within the interstitial space and dilutes the urine. Some of the sodium diffuses into the descending limb and is recycled, resulting in a countercurrent multiplication of the osmotic gradient that makes the interstitial space progressively more concentrated toward the bend of the loop (*287*). Ascending limb cells do not contain an abundance of mitochondria (*200, 227*) and the oxygen tension is low in this region. Energy for the sodium pump is probably provided by anaerobic glycolysis.

## Collecting Duct

The collecting duct has an important role in the production of a concentrated urine. Under the influence of antidiuretic hormone (vasopressin) the membranes facing the collecting duct lumen become permeable. Water then moves down the osmotic gradient into the interstitial space, which has been made considerably hyperosmotic because of the operation of the countercurrent multiplier in the loops of Henle. Water entering the cells across the apical plasma membrane (apm) leaves the cells both across basal (a) and lateral (b) membranes (*96*). This water movement renders the urine progressively more concentrated as the fluid moves through the collecting ducts, and a hyperosmotic final urine is formed. When vasopressin is absent (diuresis) the distal tubule and collecting ducts are less permeable to water, so that the dilute fluid in the distal tubule passes through the collecting duct with little concentration change and a hypoosmotic final urine is formed (*13, 305*). Under certain conditions urea is absorbed from the collecting ducts, and this could be accomplished by entrainment (p. 5).

Mammalian kidney loop of Henle and collecting duct

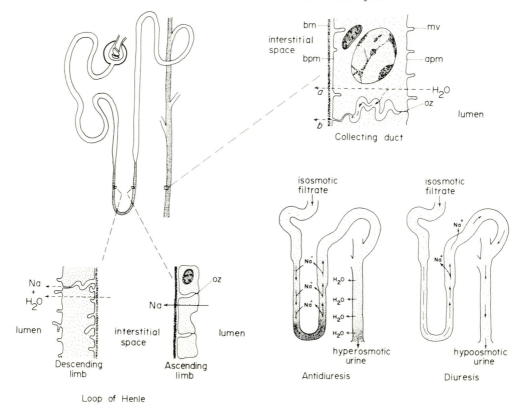

Loop of Henle

**Fig. 49.** Descending thin limb of the loop of Henle in the rabbit kidney. ×45,000. From *Z. Zellforsch. Mikrosk. Anat.* 93, 516–524. Courtesy of Dr. S. Jane Darnton.

**Fig. 50.** Ascending thin limb of the loop of Henle in the rabbit kidney. ×14,000. From *Z. Zellforsch. Mikrosk. Anat.* 93, 516–524. Courtesy of Dr. S. Jane Darnton.

**Fig. 51.** The collecting duct of the rat kidney. ×6690. Courtesy of Dr. Ruth Ellen Bulger.

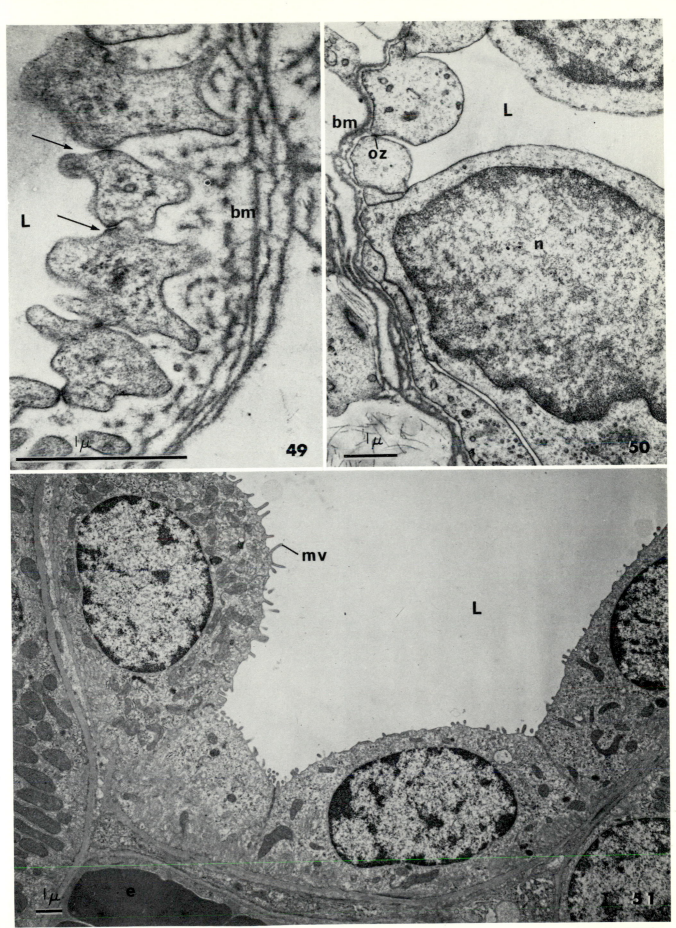

## Aglomerular Kidney of Marine Teleosts

The blood of most marine fish is less concentrated than seawater, so water leaves the body by osmosis and salt enters by diffusion. Water loss across the body surface is replaced by drinking seawater and the excess salts are excreted by the gills (p. 32). The kidneys have a relatively minor role in osmoregulation. Mechanisms for filtering and diluting the urine are not adaptive, and in various marine fish the glomerular and distal tubular functions are reduced or absent (*261*). For example, in the goose or angler fish *Lophius*, each kidney has only about seventy-eight glomeruli for the whole population of 150,000 tubules, and these are "pseudoglomeruli" since they do not connect with the functional tubules (*177*). Other species, such as the toadfish *Opsanus* have no glomeruli.

In aglomerular fish the main functional unit of the kidney is a simple blind-ended tubule that drains into a collecting duct. The tubule is highly convoluted, but there is no uniformity in the shape of the convolutions. The cells closely resemble those of the mammalian proximal tubule, as they have elaborate basal interdigitations (bi) associated with mitochondria, a typical brush border (mv), and many heterophagic vacuoles (*29, 76, 192, 193*).

The blood supply to the aglomerular kidney is under such a low hydrostatic pressure that ultrafiltration in the usual sense cannot occur. Osmotic filtration (Fig. 5) may be the mechanism by which secretion is accomplished. Aglomerular tubules thus appear to have functional as well as structural similarities with insect Malpighian tubules (p. 18). The structure of the collecting ducts is simple, indicating that they have a minor role in urine formation.

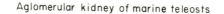

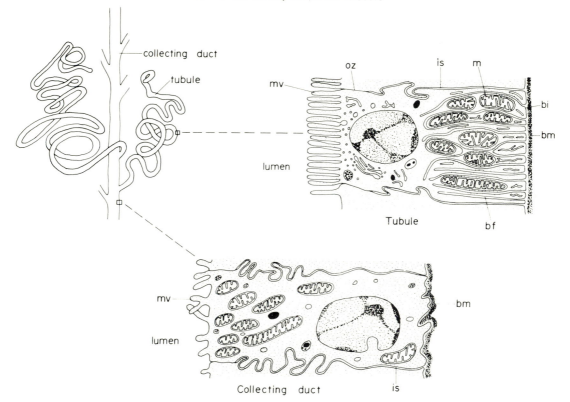

**Fig. 52.** The whole secretory tubule of the aglomerular goose fish *Lophius piscatorius*. ×6000. From *Z. Zellforsch. Mikrosk. Anat.* **104**, 240–258. Courtesy of Dr. Jan L. E. Ericsson.

**Fig. 53.** The apical microvilli of aglomerular tubule, showing strands of glycocalyx between them. ×40,000. From *Z. Zellforsch. Mikrosk. Anat.* **104**, 240–258. Courtesy of Dr. Jan L. E. Ericsson.

**Fig. 54.** Basal infoldings of the aglomerular tubule of the goose fish. ×24,500. From *Z. Zellforsch. Mikrosk. Anat.* **104**, 240–258. Courtesy of of Dr. Jan L. E. Ericsson.

**Fig. 55.** The lateral plasma membrane. ×21,000. From *Z. Zellforsch. Mikrosk. Anat.* **104**, 240–258. Courtesy of Dr. Jan L. E. Ericsson.

**Fig. 56.** The apical surface of the aglomerular tubule. From *Z. Zellforsch. Mikrosk. Anat.* **104**, 240–258. ×35,000. Courtesy of Dr. Jan. L. E. Ericsson.

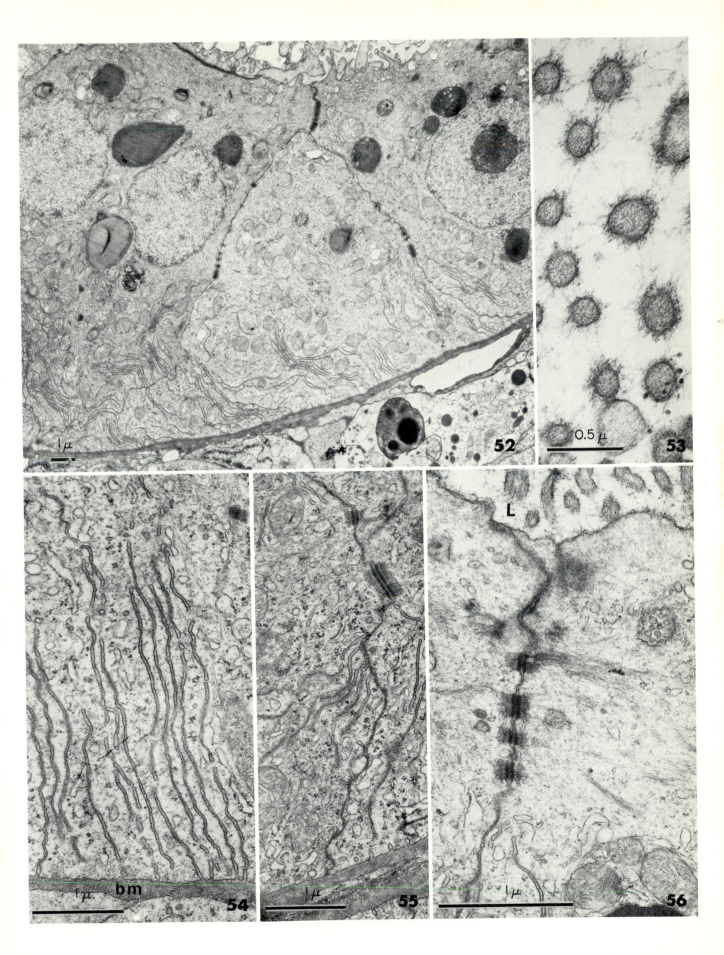

52

0.5 μ    53

1 μ

1 μ    bm    54

1 μ    55

L

1 μ    56

## Reptilian Kidney

In reptiles nitrogenous metabolism yields insoluble uric acid which can be excreted in crystalline form with minimal water loss. The rate of filtration is reduced, and there is a reduced vascular supply to the glomeruli. Most of the filtered water is reabsorbed. The distal tubule reabsorbs salt in excess of water so that the final urine is hypoosmotic to the blood. There is further modification of the urine in the cloaca (242). Much of the osmotic pressure of the final urine is accounted for by ammonia and bicarbonate ions. The urine is never hyperosmotic to the blood, but the volume of urine is quite low. Marine reptiles have salt glands (p. 28) that remove excess salt.

The proximal (Fig. 57) and distal tubules (Fig. 58) are different from those of mammals in that they lack the extensive basal interdigitations (53). Instead, they have extensive intercellular spaces (is) which distend during fluid uptake (240).

The proximal tubule reabsorbs water and sodium chloride in isosmotic proportions, and is functionally analogous to gallbladder and mammalian proximal tubule, but only structurally analogous to gallbladder (p. 52). The distal tubule cells resemble the gallbladder even more closely, but reabsorb ions in excess of water so that the final urine is hypoosmotic to the blood. The formation of an absorbate that is more concentrated than the fluid in the lumen may be accomplished by local osmosis in which equilibration is not complete due to the extra width of the channels, or due to secretion of ions along the whole length of the channels rather than just towards their closed ends (240).

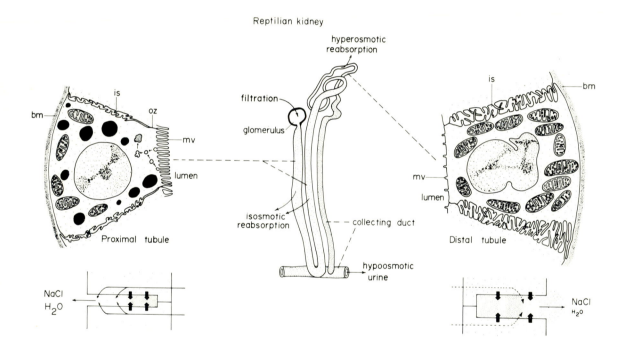

**Fig. 57.** The proximal tubule of the kidney of a Northern Blue tongue lizard *Tiligua scincoides*. ×8600. Courtesy of Dr. Lowell E. Davis.

**Fig. 58.** Distal tubule of the lizard kidney. ×8600. Courtesy of Dr. Lowell E. Davis.

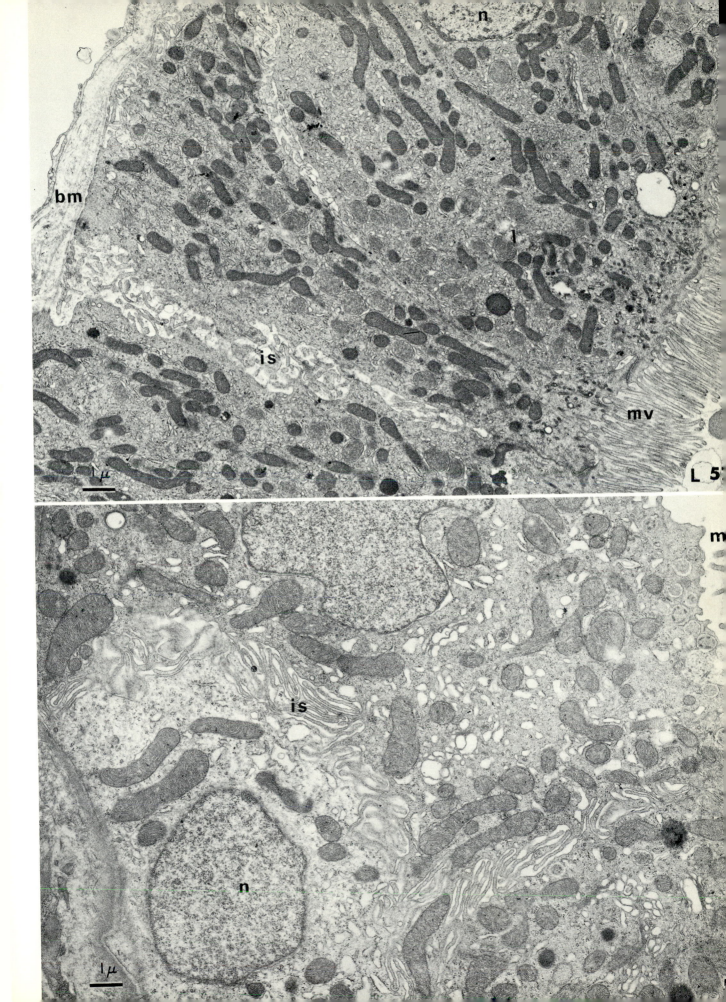

## DIGESTION AND ABSORPTION

Animals possess a range of epithelia involved in ingestion, digestion, and absorption of food. Salivary glands provide certain digestive enzymes and also lubricate the food during ingestion. Once food enters the intestine, enzymes are added to hydrolyze proteins, carbohydrates, and fats into their component subunits. These molecules, along with ions and water, are then absorbed by specialized regions of the intestine. Some of the digestive secretions come from the gut wall (e.g., from the gastric glands of the stomach) while others originate elsewhere (e.g., liver and pancreas).

Many of the epithelia concerned with digestion and absorption have striking structural similarities that most probably reflect functional similarities. For example, the apical absorptive surfaces of the vertebrate small intestine and insect midgut are almost identical. The parietal cells of vertebrate stomach also closely resemble the secretory cells of insect salivary glands. Finally, the rumen resembles the frog skin (described in the previous section) and probably absorbs sodium by a similar mechanism.

### Mammalian Salivary Gland

The salivary glands of mammals produce a watery secretion that moistens and prepares the food for ingestion and digestion. The three main glands are the parotid (serous), sublingual (mucous), and submaxillary or mandibular (serous and mucous). Despite differences in the nature of their secretions, their general organization and membrane architecture are very similar (116, 155, 205, 269–271, 304).

Bulbous acini are connected to striated ducts by short thin intercalated ducts composed of small unspecialized cells. The acini have pyramidal cells arranged around a narrow central lumen (Fig. 63). The apical membrane has a few short microvilli (mv) and is often arranged to form canaliculi (*) which open into the lumen (Fig. 60). Occluding zonules (oz) and desmosomes (d) connect adjacent cells in this apical zone. The remainder of the lateral membrane is somewhat folded and adjacent cells interdigitate. In some species the basal plasma membrane is also infolded and neighboring cells interdigitate (Fig. 61). Each acinus is usually associated with a single stellate myoepithelial cell (mc), while longitudinally arranged myoepithelial cells lie along the intercalated ducts. Myoepithelial contractions squeeze the acini and cause a shortening and widening of the intercalated ducts to allow fluid to pass into the striated ducts (268).

The basal membrane of striated duct cells (Fig. 62) is highly folded by a complex series of interdigitations with neighboring cells (heavy stippling represents projections from neighboring cells). The lateral plasma membrane also has interdigitations. The apical plasma membrane usually has a few short microvilli (Fig. 59, mv). Secretory granules (g) of various sizes and number are found in the striated duct cells of most glands.

Stimulation of submaxillary and parotid glands results in formation of a hypoosmotic saliva which originates as an isosmotic secretion in the acini and is made less concentrated by reabsorption of ions in the striated ducts (176, 311). The intercellular canaliculi between acinar cells may be the sites of osmotic equilibration during isosmotic fluid secretion. The striated duct cells resemble other cells engaged in hyperosmotic absorption such as distal kidney tubule (p. 36) and segmental gland tubule (p. 14).

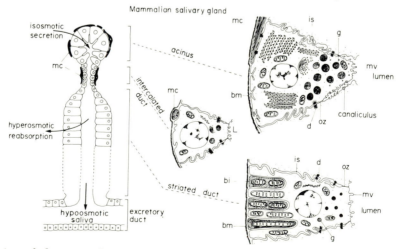

Mammalian salivary gland

Fig. 59. Apical region of the striated duct of human submaxillary salivary gland. ×19,000. Courtesy of Dr. Bernard Tandler.

Fig. 60. A canaliculus between adjacent acinar cells of human labial salivary gland. ×25,000. Courtesy of Dr. Bernard Tandler.

Fig. 61. Basal surface of the acinar cell of human submaxillary salivary gland. ×45,000. Courtesy of Dr. Bernard Tandler.

Fig. 62. Basal region of human submaxillary striated duct. ×28,000. Courtesy of Dr. Bernard Tandler.

Fig. 63. Acinar serous cell of human submaxillary gland. ×10,500. Courtesy of Dr. Bernard Tandler.

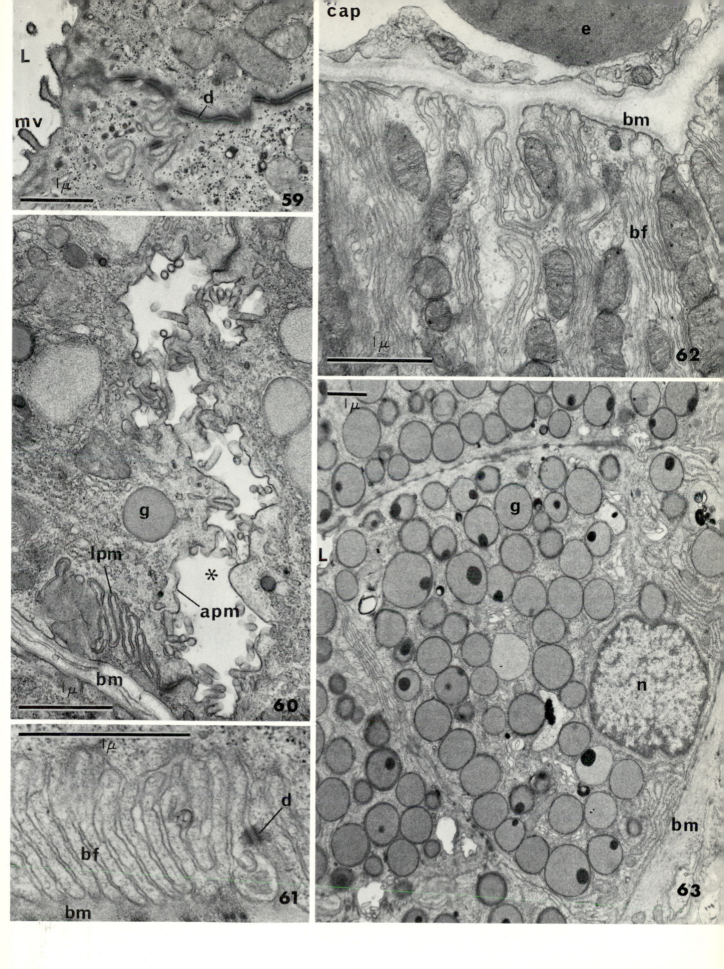

## Insect Salivary Gland

The salivary glands of most insects produce a watery saliva containing digestive enzymes. Studies on the mechanism of fluid secretion have concentrated on two species, a dipteran, *Calliphora,* and an orthopteran, *Periplaneta.*

### Diptera

The salivary glands of the adult blowfly *Calliphora* consist of a pair of long tubes extending down the length of the body. Most of the tube has a single cell type (stippling) characterized by large secretory canaliculi (∗) formed by invaginations of the apical plasma membrane (Fig. 65) (197). Parallel leaf-like microvilli (mv) line both the canaliculi and the free apical surface. The lateral plasma membrane is straight and has a septate desmosome (sd) in the apical region. The basal plasma membrane has numerous infoldings (bf) closely associated with mitochondria. In addition to large mitochondria, the cytoplasm also has granules which may contain amylase. A short segment of the tube (clear region) has much smaller cells which lack secretory canaliculi and have short wide basal infoldings closely associated with mitochondria.

Secretory cells are stimulated by 5-hydroxy-tryptamine (acting via cyclic AMP) to produce an isosmotic potassium-rich fluid (15, 197). The basal infoldings and long canaliculi may function in isosmotic fluid flow by creating standing osmotic gradients as postulated by Diamond and Bossert (57). The short absorptive segment modifies the saliva by absorbing potassium, so that the final saliva is hypoosmotic to the blood (197).

### Orthoptera

The racemose glands of Orthoptera (e.g., the cockroach *Periplaneta*) are confined to the thorax. The acini have different cell types (Fig. 66) (138, 141). One cell type has canaliculi (∗) and thus resembles the secretory cells in *Calliphora.* The remaining acinar cells may be mucous, zymogenic, or both. They have extensive endoplasmic reticulum (rer) and numerous large granules (g). Duct cells (Fig. 64) have long basal infoldings associated with mitochondria and a folded apical membrane (af). The duct is lined with cuticle (c).

The structure of the gland, which closely resembles that in mammals (previous section) suggests that the acini produce a primary secretion which is modified as it passes through the ducts, as in *Calliphora.*

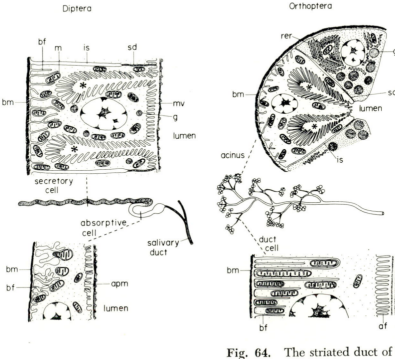

Insect salivary gland

Fig. 64. The striated duct of the salivary gland of the cockroach. ×14,850.

Fig. 65. Survey picture of the secretory region of the salivary gland of *Calliphora erythrocephala.* ×12,150.

Fig. 66. Acinar cell of the salivary gland of the cockroach, *Periplaneta americana.* ×5850.

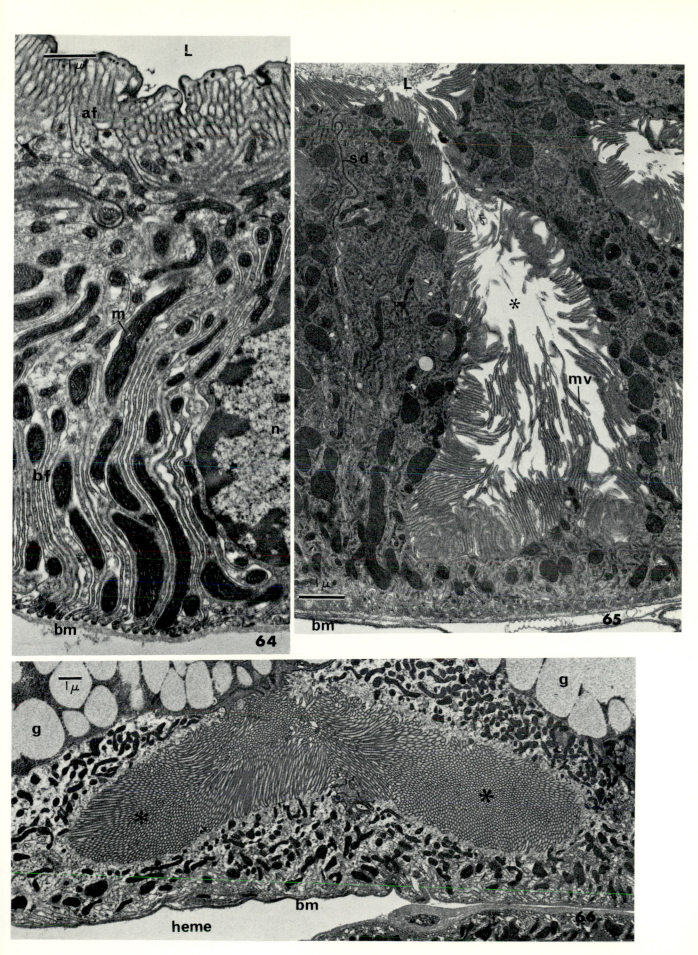

## Parietal Cell of Vertebrate Stomach

Gastric glands in the vertebrate stomach secrete gastric juice containing enzymes (pepsin and renin) dissolved in a dilute hydrochloric acid solution (0.04–0.05%). The glands are branched tubes communicating with the stomach through gastric pits. In mammals the HCl-secreting parietal cells are usually found in the middle of the gland, whereas enzymes are secreted by chief cells at the base. In amphibia, both processes are probably carried out by the single cell type which makes up the gastric gland.

### Amphibia

The amphibian parietal cell has an extensive interconnected tubular system (Fig. 67), which occupies most of the apical region of the cell and may occasionally communicate (arrows) with the apical cell surface (120, 249, 251, 253). Earlier studies had revealed the presence of a vesicular system, but this was probably an artifact caused by breakdown of the tubules during fixation. During HCl secretion the surface area of the apical membrane is probably increased by fusion with the tubular system (250, 291). The apical

and tubular membranes have similar dimensions and staining properties (163, 236,). The tubular elements may first fuse with the apical membrane and then evert to form microvilli (252, 253). The lateral membrane of adjacent cells is infolded and interdigitated to produce a complex intercellular compartment (is) open on the blood side but closed to the lumen by occluding zonules (oz). Zymogen granules (g) and mitochondria (located mainly in the basal zone) occupy the remainder of the cell.

### Mammals

The mammalian parietal cell differs from the amphibian type because it lacks zymogen granules and has extensive secretory canaliculi (∗) opening into the lumen (Fig. 68) (110, 120, 122, 162, 233, 255, 264). These canaliculi, which are lined with microvilli, extend deep into the cell so that they are often closely associated with the irregular basal infolds. Most mammalian parietal cells also have an extensive tubular system (254), but there is no connection with the lumen (119). The secretory canaliculi may function in isosmotic HCl secretion by providing long narrow channels where water movement can be coupled to the active extrusion of hydrogen and chloride from the cell.

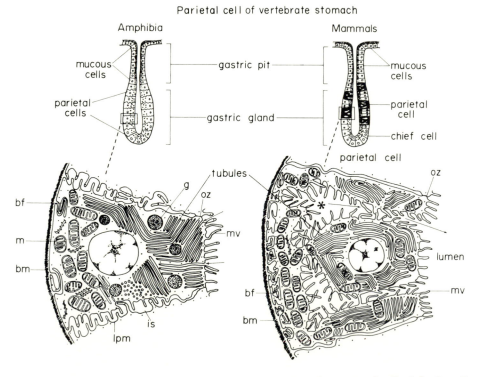

Fig. 67.   The parietal cell of the frog *Rana castebeiana.* ×25,200. Courtesy of Dr. Albert W. Sedar.

Fig. 68.   The parietal cell of the bat. ×10,000. From Bloom and Fawcett "A Textbook of Histology," Saunders, Philadelphia. Courtesy of Dr. Susumu Ito.

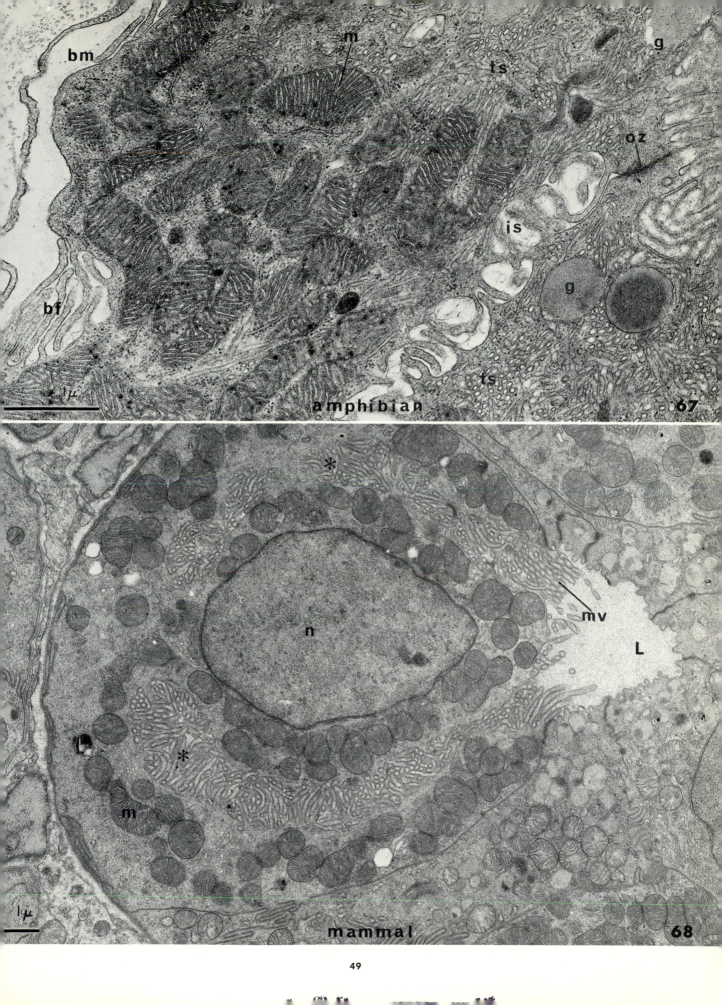

amphibian

67

mammal

68

## Vertebrate Liver

The liver is the largest gland in the body and has a multitude of functions in secretion, metabolism, and storage. It has a special position in the circulation, interposed between the intestinal tract and the general circulation. Most of the material absorbed from the gut is processed by the liver. The liver secretes an isosmotic fluid, bile, that contains bile salts and acids, cholesterol, bilirubin, fatty acids, and electrolytes. Bile also contains the harmless products or conjugates of toxic substances such as drugs that have been detoxified by the hepatic cells.

Hepatic parenchymal cells (Fig. 69) are arranged in single-layered sheets (hepatic cell plates) lining a modified capillary bed, the hepatic sinusoids (32, 151, 171, 309). The sinusoids (Fig. 69, sin) are lined with ordinary capillary endothelial cells (end) and reticuloendothelial (Kupffer) cells (Fig. 69, KC). The endothelial cells rest on projections from the hepatic cells to form the space of Disse (Dis). This space is in free communication with the capillary lumen through large gaps (arrows) between the endothelial cells (21, 32, 108). Kupffer cells remove particulate matter from the blood by phagocytosis. The apical surface of the hepatic cells faces the bile canaliculi (∗) while the

basal surface faces the sinusoid. Basement membrane is absent along most of the length of the sinusoid (32) except calf liver, which has a distinct basement membrane (309). The narrow intercellular space between the hepatic cells is in free communication with the space of Disse, but is closed off from the canaliculi by occluding zonules (Fig. 70, oz). Gap junctions (om) occur as plaques along the intercellular cleft, but do not hinder the movement of tracers such as peroxidase from the space of Disse up to the occluding zonules adjacent to the canaliculi (94, 178). The canaliculi form an extensive intercellular network draining into larger bile ducts which merge to form the hepatic duct. The latter conveys bile to the gallbladder. Liver cells contain a store of glycogen and have a well-developed rough endoplasmic reticulum involved in the synthesis of plasma proteins and smooth reticulum engaged in the synthesis of drug hydroxylating enzymes (130, 195).

The precise mechanism of isosmotic bile secretion is unknown. By analogy with other systems, standing gradients might be created within the bile canaliculi due to solute transport from the hepatic cells. Water would then follow passively, carrying other solutes by solvent drag.

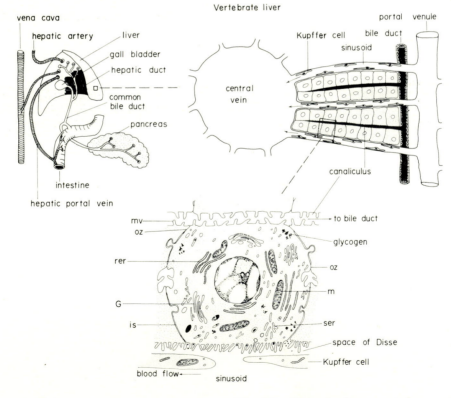

Vertebrate liver

**Fig. 69.** Survey view of the hepatic parenchymal cell of a rat. ×7000. From Ashford and Dowben "General Physiology," Harper & Row, New York. Courtesy of Drs. Thomas P. Ashford and Robert M. Dowben.

**Fig. 70.** Lateral plasma membrane of a mouse liver cell, showing occluding zonule and gap junction. ×32,500. *Inset,* Detail of the gap junction. ×420,000. From *J. Cell Biol.* **45,** 272–290. Courtesy of Drs. Daniel A. Goodenough and Jean-Paul Revel.

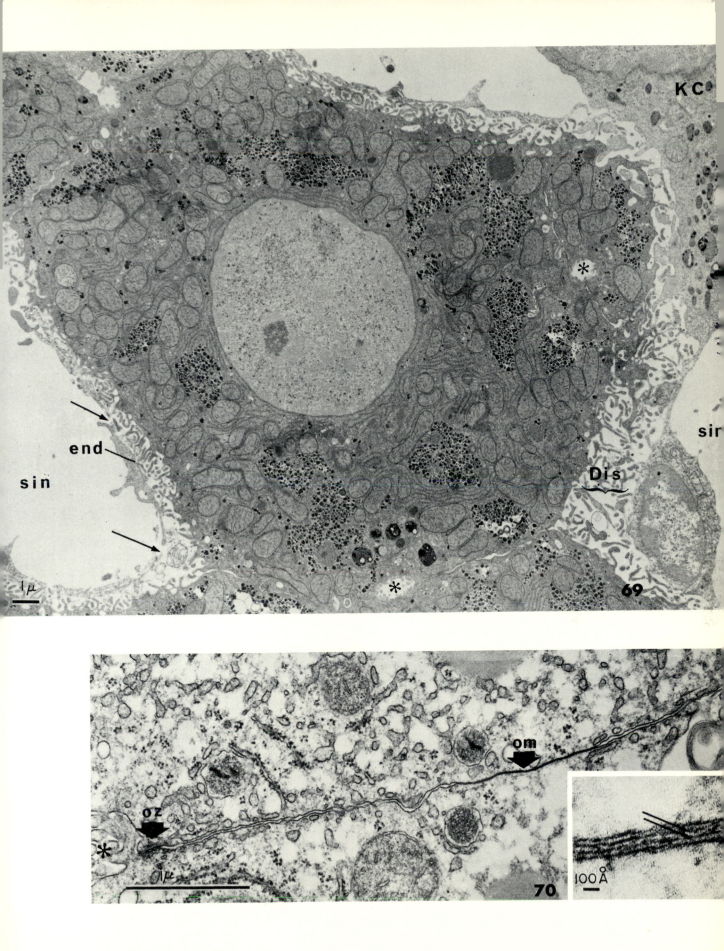

## Vertebrate Gallbladder

The gallbladder concentrates and stores bile (secreted by the parenchymal cells of the liver, previous section) by an isosmotic absorption of sodium chloride and bicarbonate resulting in a six- to tenfold concentration of bile pigments and salts. The concentrated bile flows through the common bile duct into the duodenum.

The columnar epithelium (Fig. 71), which lies on a lamina propria, has a similar structure in different vertebrates (40, 79, 107, 128, 136, 277). The apical plasma membrane has short (0.5–4 μ) microvilli (Fig. 73, mv) covered with a mucopolysaccharide coat. The lateral membranes fuse to form occluding zonules, followed by adhering zonules (az). The remainder of the lateral membrane forms a complex series of projections which interdigitate with neighboring cells. The width of the intercellular space (is) varies (see below) but the basal portion of the channel is always closed down to a constant width of 200–500 Å (Fig. 72,

arrow). The basal plasma membrane is flat and there is a thin basement membrane (bm). Mitochondria tend to be concentrated in the apical region, but not in the fibrous terminal web.

Studies of the structure and function of the gallbladder have led to the development of the standing gradient model which may be applicable to many other epithelia (55, 56, 58, 59, 62, 136, 277). Fluid absorption from the lumen depends on an electrically neutral pump which transports sodium and chloride into the intercellular channels. This creates a local osmotic gradient causing a passive flow of water. The solute concentration within the long narrow intercellular channels may vary along the channel length (stippling) if solute input (large arrows) is restricted to the closed end, resulting in a standing osmotic gradient. The hyperosmotic fluid at the closed end will gradually be diluted by water entering along the length of the channel (broken arrows) so that the isosmotic fluid finally leaves the basal end of the channel. The width of the intercellular channel has been correlated with the rate of fluid movement (136, 277).

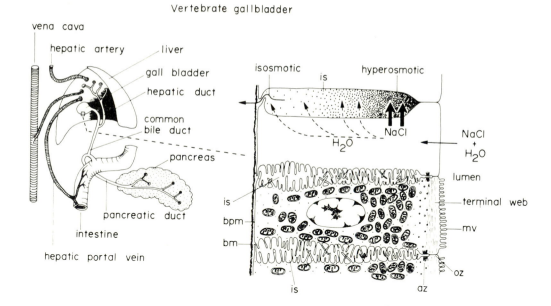

Vertebrate gallbladder

Fig. 71.  Rabbit gallbladder fixed while transporting fluid *in vitro*. ×3900. From *J. Gen. Physiol.* **50,** 2031–2060. Courtesy of Dr. John McD. Tormey.

Fig. 72.  Basal surface of a rabbit gallbladder fixed while absorbing water at a rate of 960 μliter/hour *in vivo*. ×18,000. From *J. Cell Biol.* **30,** 237–268. Courtesy of Dr. Gordon I. Kaye.

Fig. 73.  Apical surface of rabbit gallbladder. ×25,800. From *J. Cell Biol.* **30,** 237–268. Courtesy of Dr. Gordon I. Kaye.

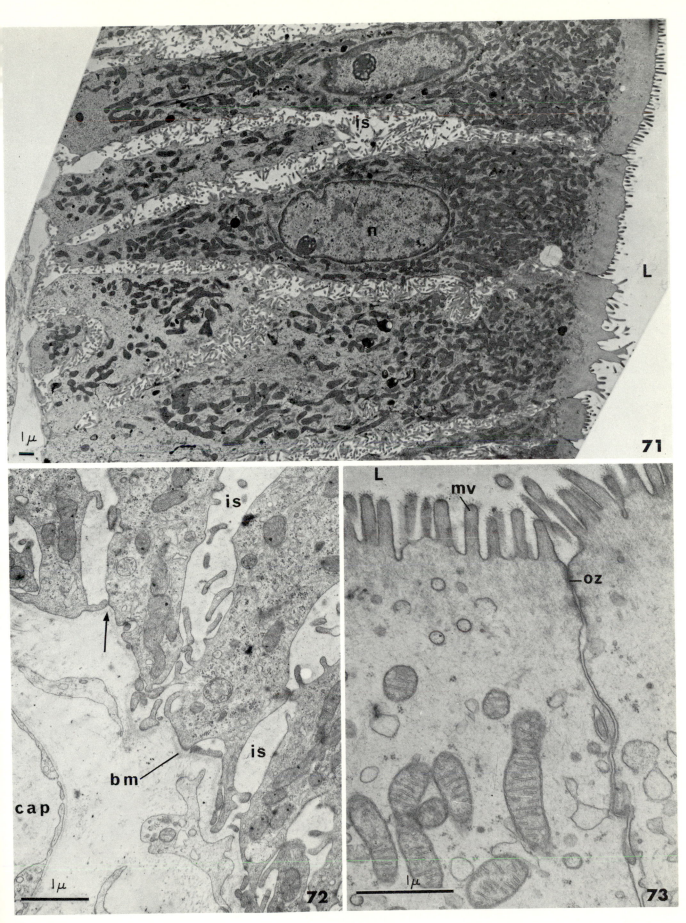

## Vertebrate Exocrine Pancreas

The exocrine pancreas secretes a clear isosmotic fluid that flows into the duodenum. Pancreatic juice contains a variety of enzymes, including amylase, lipase, and several proteases. The fluid is also rich in bicarbonate, which serves to neutralize the highly acidic material entering the duodenum from the stomach.

The exocrine pancreas consists of acini and intercalated ducts. The acini are highly irregular in form as they contain a mixed population of acinar and centro-acinar cells (Fig. 74). The latter do not completely line the acinar lumen. Instead the arrangement is such that sections through an acinus usually show a lumen bounded by both cell types (71). Yet a portion of each acinar cell either faces the central lumen or has access to it via the canaliculi (*). The acinar cells contain an abundant rough endoplasmic reticulum (rer) and Golgi complex (G) involved in production of zymogen granules (g). Zymogen consists of inactive proteases which become active when they reach the intestine, where specific activators are secreted. Centro-acinar cells lack canaliculi but their lateral surfaces

are somewhat folded, particularly near the basal surfaces [Figs. 74 (arrows), 76]. Both cell types have short microvilli (mv) and in the human (illustrated here) each centro-acinar cell has a cilium (Fig. 75, cil) (139). The intercalated ducts are also lined by centro-acinar cells, and possibly by another type of cell (71).

The pancreas secretes an isosmotic fluid at all flow rates. The juice contains $Na^+$ and $K^+$ at the same concentrations as plasma. Anionic composition depends on rate of secretion. The bicarbonate concentration increases as flow rate increases, and chloride concentration decreases so that the sum of these two anions remains relatively constant. Secretin stimulates fluid flow without affecting enzyme output, while pancreozymin does not affect flow rate but causes depletion of zymogen granules from acinar cells. Active bicarbonate secretion by the centro-acinar cells may bring about passive flow of water and other ions into the lumen (126). This fluid is modified as it flows along the ducts, with bicarbonate in the lumen exchanged for chloride in the blood. At lower flow rates there is more opportunity for exchange and the secreted fluid is rich in chloride (37, 126).

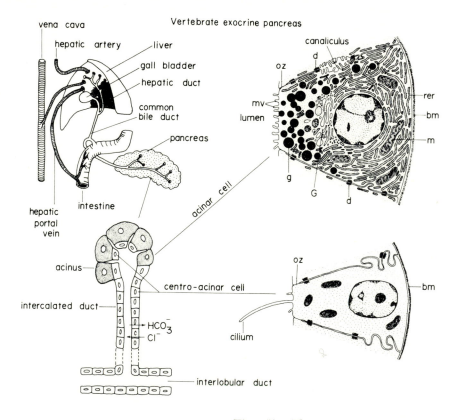

Vertebrate exocrine pancreas

**Fig. 74.** Survey of an acinus of human pancreas, showing acinar and centro-acinar cells. ×7800. From Z. *Zellforsch. Mikrosk. Anat.* **113,** 322–343. Courtesy of Dr. Horst Kern.

**Fig. 75.** The acinar lumen, including cilia of centro-acinar cells. ×11,800. From Z. *Zellforsch. Mikrosk. Anat.* **113,** 322–343. Courtesy of Dr. Horst Kern.

**Fig. 76.** Details of the junction between two centro-acinar cells in the intercalated duct. ×60,000. Courtesy of Dr. Horst Kern.

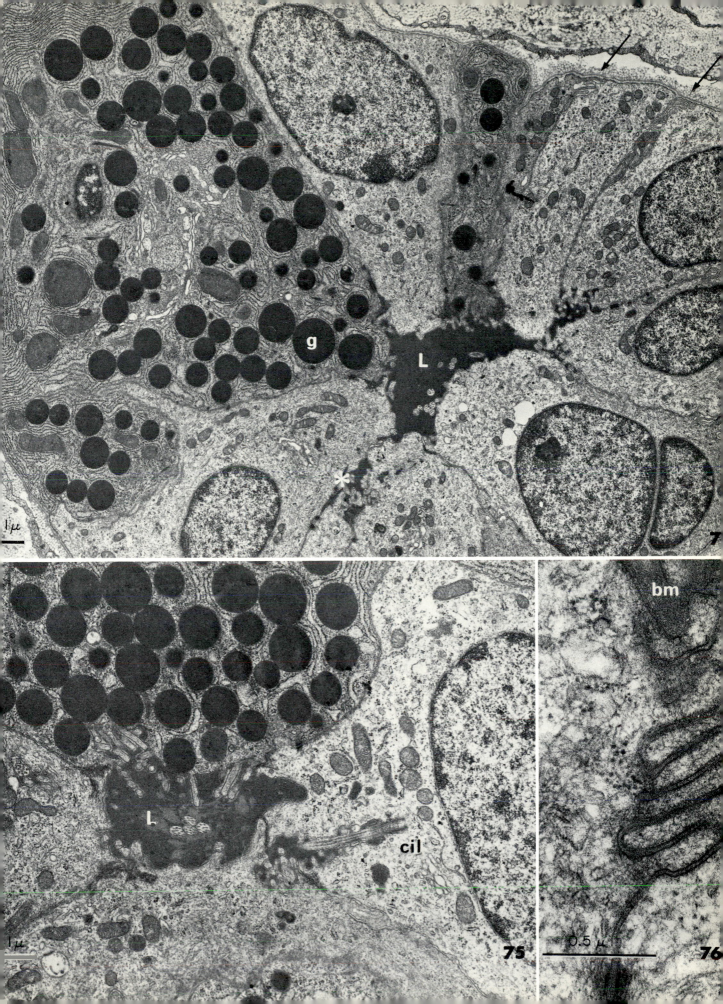

g

L

*

74

1μ

L

cil

75

1μ

bm

0.5μ

76

## Vertebrate Small Intestine

The intestine absorbs ions and water together with a wide range of nutrients. A characteristic feature of the absorptive surface is the enormous increase in area achieved through the Valves of Kerckring (folding of whole mucosa), villi, and microvilli. The crypts of Lieberkühn, composed of goblet, enterochromaffin, Paneth, and undifferentiated cells (281) are mainly involved in cellular proliferation and secretion.

The structural organization of the absorptive cells (Fig. 77) is similar in different vertebrates and in the different regions of the small intestine (202, 282, 283). The large columnar cells have an elaborate brush border covered with a glycocalyx (121). Within each microvillus (mv) is a core of fine filaments which are extensions of the filamentous meshwork of the terminal web (Fig. 78). This fibrillar assembly probably stabilizes the microvilli, thus maintaining a large absorptive surface. The area of the lateral membrane (lpm) is increased by interdigitation with neighboring cells but the basal membrane (bpm) is flat. The lateral plasma membranes are closely apposed near the basal surface

to form a narrow channel (arrows, Fig. 79) similar to that found in gallbladder (p. 52), toad bladder (p. 26), and kidney collecting duct (p. 38).

Absorption of water and nonelectrolytes are closely linked to sodium transport. Sodium entering passively across the apical membrane (a and b) is actively extruded (solid arrow) into the intercellular space (89, 247) to create the osmotic gradient that brings about a passive flow of water (49) resulting in the formation of an isosmotic absorbate (c). Specific sodium-dependent carrier mechanisms (a and b) located on the apical membrane absorb sugars and amino acids (47) which then leave the cell by diffusing either across the basal membrane (e) or via the intercellular pathway (d) where the simultaneous fluid flow (c) exerts a continuous solvent drag. Certain aspects of lipid absorption can be visualized in the electron microscope (36, 203, 217, 267). Lipids degraded to fatty acids and monoglycerides in the lumen (f) are absorbed (g) and then reconstituted within the smooth ER (h) to form lipid droplets. Lipid droplets released into the intercellular space (i) are termed chylomicra, and move across the basement membrane (j) to enter the lacteals, or blindly ending lymphatic vessels.

Vertebrate small intestine

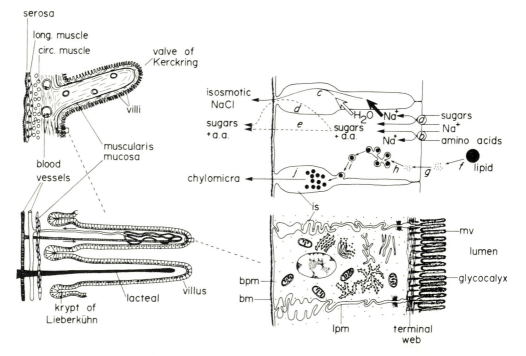

Fig. 77.  Columnar cells from the intestine of the guinea pig. A portion of a goblet cell appears at left. ×6200. From *J. Cell Biol.* 34, 123–155. Courtesy of Dr. Keith R. Porter.

Fig. 78.  Apical surface of the intestinal columnar cell. ×32,000. From *J. Cell Biol.* 34, 123–155. Courtesy of Dr. Keith R. Porter.

Fig. 79.  Basal surface of the intestinal cell. ×23,000. Courtesy of Dr. Keith R. Porter.

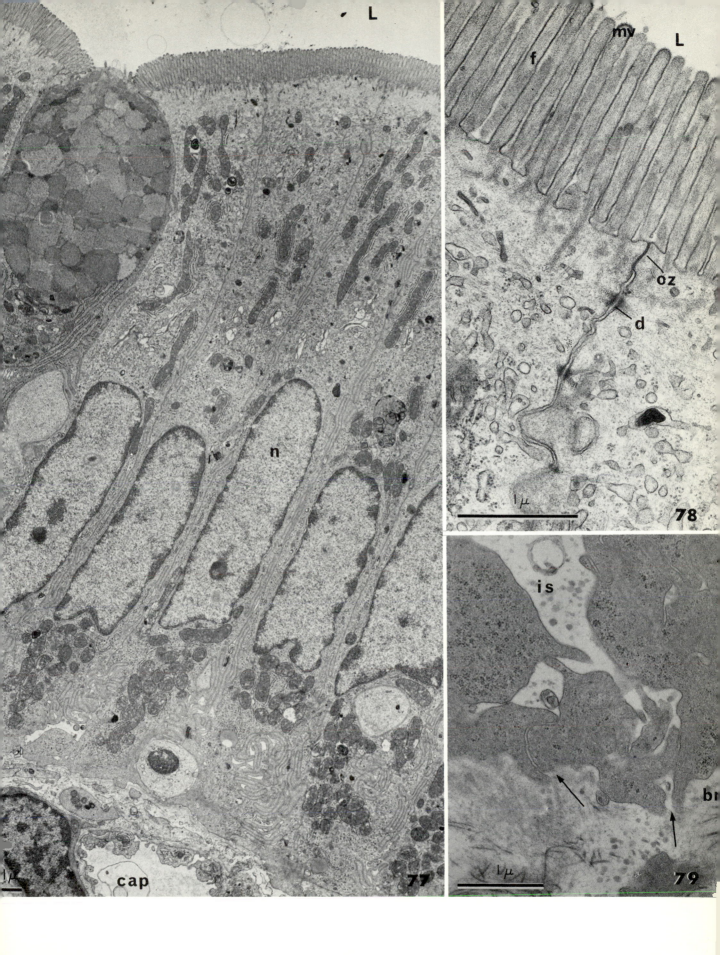

## Insect Midgut

The insect midgut or ventriculus, like the intestine of vertebrates, is the major site of absorption of water, ions, and nutrients. In addition there is secretion of dyes and perhaps fluid into the midgut lumen (*220, 302*).

The general structure of the columnar absorptive cells (Fig. 80) of insect midgut is similar in the various insects that have been studied (*95, 131, 188, 189, 262*), and is also similar to that of vertebrate intestine (previous section). The principal difference between a vertebrate and an insect absorptive cell is that the latter has highly folded basal and lateral membranes. The foldings form an extensive compartment that is closed off from the hemolymph except for a few narrow openings (Fig. 81) (*14*). Mitochondria are closely associated with the infolded membranes. The cells have an abundance of rough endoplasmic reticulum, Golgi complexes, and small lipid droplets. The latter are the only organelles present in the terminal web region. The microvilli are neatly arranged in a hexagonal array, and fine filaments of glycocalyx extend between them (Fig. 82) (*196*).

The endoplasmic reticulum probably synthesizes digestive enzymes that are secreted into the lumen, and may also synthesize some blood proteins (*235*). The major properties of the midgut columnar cells can be summarized by a model that is similar to that originally proposed to account for sodium uptake by frog skin (*194*). Sodium pumped from the cell into the hemolymph is replaced by sodium entering from the lumen because the apical membrane is more permeable. Sodium uptake generates a flow of water that drags other solutes along, while concentrating impermeant molecules in the lumen. This creates a gradient favoring the passive diffusion of these molecules from the lumen to the blood. Treherne (*278, 280*) suggests that this mechanism may account for amino acid uptake. He also finds that monosaccharides are absorbed by a facilitated diffusion in which a favorable gradient between lumen and hemolymph is maintained by converting monosaccharides into trehalose as soon as they enter the hemolymph. In insects with a high blood glucose concentration, glucose may be absorbed by a mechanism similar to that of amino acids. Alternatively, these solutes may be absorbed by entrainment (see Fig. 6).

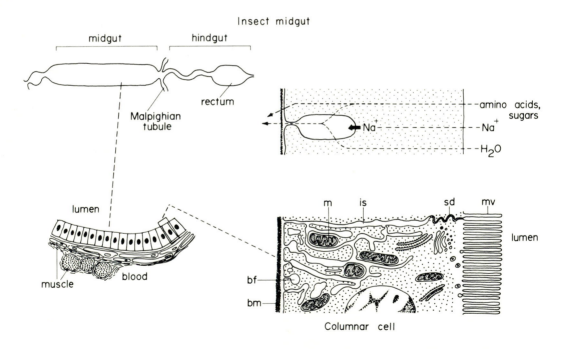

Insect midgut

Fig. 80.  Survey view of the whole epithelium of the midgut of a cockroach *Periplaneta americana*. ×4200.

Fig. 81.  High magnification of the basal region of cockroach midgut. ×17,500.

Fig. 82.  Apical microvilli, showing strands of glycocalyx between them. ×80,000.

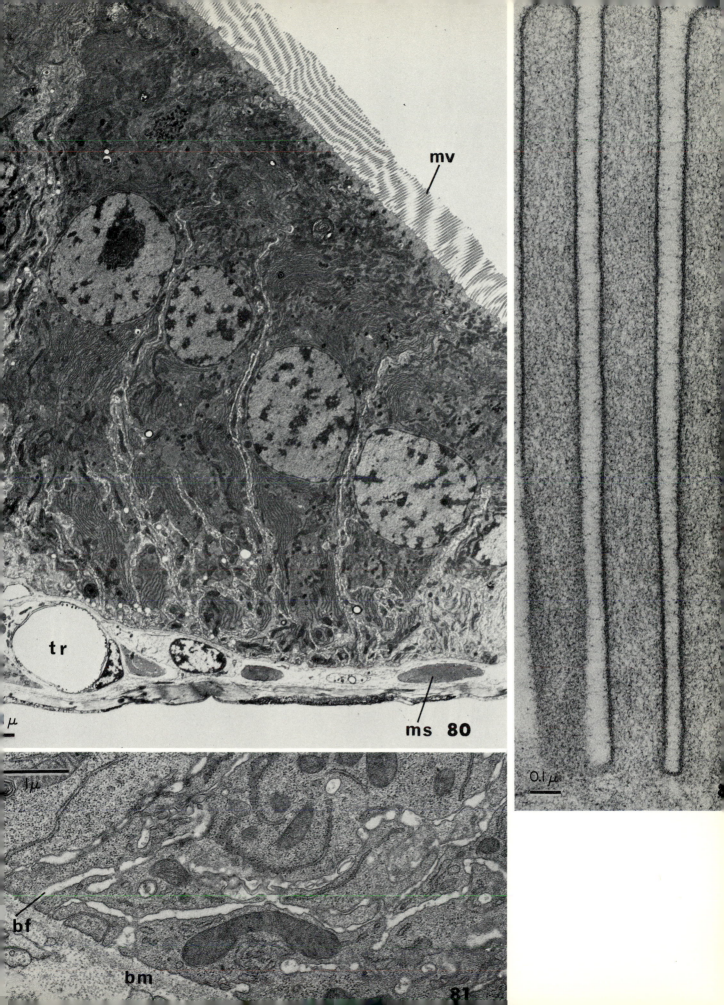

mv

tr

$\mu$

ms 80

$1\mu$

bf

bm

81

$0.1\mu$

8

## Rumen

Ruminant artiodactyls (cow, sheep, goat, ox, llama, camel) possess an elaborate esophageal-stomach complex that serves as a storage reservoir for food materials and symbiotic microorganisms. Food is ingested (a) and stored in the rumen and reticulum for long periods. Periodically a portion is returned to the mouth for further chewing, which breaks up the food and allows better access for the digestive enzymes produced by the symbionts. The "cud" is returned to the rumen (b) with large volumes of saliva. The large amounts of sodium in the saliva are recovered by active absorption from the rumen. Water and chloride are probably absorbed as well. Sugars produced by breakdown of plant material including cellulose are not absorbed by the animal but instead are utilized by the symbiotic microorganisms. Anaerobic glycolysis results in production of volatile fatty acids, including acetic, propionic, butyric, and some valeric. These are absorbed by the epithelium and converted to glucose, and provide 70–80% of the total energy intake of the animal. Proteins are also broken down by the symbionts and some amino acids are absorbed by the rumen epithelium. However, most of the amino acids are utilized by the micro-organisms and are not available to the host until the food enters the abomasum or stomach, where normal digestion proceeds. The symbionts can also synthesize proteins from ammonia and urea that enter the rumen by diffusion from the blood and via the saliva (63, 68, 118).

The rumen is lined by a keratinizing nonglandular epithelium (Fig. 84) similar in structure to amphibian (p. 24) and other skin (112, 246). Individual cells have a relatively flat surface facing the lumen, but the surface is folded into macroscopic paddle-shaped papillae (Fig. 83) with highly vascular cores. The horny surface cells often contain large vacuoles (Fig. 84, v) (112). There are differences between rumen and frog skin. The germinative cells of the base of the rumen (basal cells) seem to form a separate layer because of their elaborately folded lateral plasma membranes (Fig. 86). Cells of the midepithelium have a dense population of mitochondria. Cells are joined by desmosomes (d) and gap junctions (Fig. 85, arrow) but not by occluding zonules (112) although tracer studies (113) show a barrier to diffusion across the epithelium at the cornified layer.

The mechanism of sodium absorption is probably similar to that in frog skin. The extent to which sodium absorption generates a flow of water is unknown.

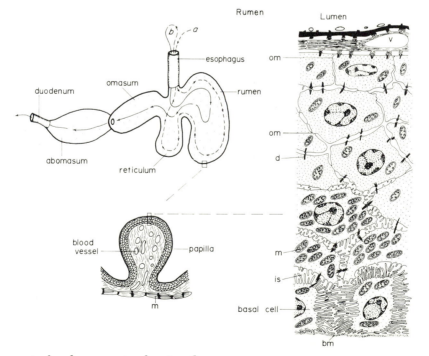

Fig. 83.   Whole mount of a sheep rumen showing the paddle-shaped papillae and smaller papillae at left. ×2.9. From *J. Ultrastruct. Res.* **30**, 385–401. Courtesy of Dr. Ray C. Henrikson.

Fig. 84.   Survey of the whole epithelium. Bacteria adhere to the apical surface. ×4750. From *J. Ultrastruct. Res.* **30**, 385–401. Courtesy of Dr. Ray C. Henrikson.

Fig. 85.   Intercellular spaces and junctions at mid-epithelium. ×66,000. From *J. Ultrastruct. Res.* **30**, 385–401. Courtesy of Dr. Ray C. Henrikson.

Fig. 86.   Highly folded membranes of the basal cells. ×28,500. From *J. Ultrastruct. Res.* **30**, 385–401. Courtesy of Dr. Ray C. Henrikson.

## Stria Vascularis of the Vertebrate Inner Ear

The stria vascularis lines the outer surface of the cochlear duct, the portion of the inner ear containing the acoustic sensory cells of the organ of Corti. The stria produces endolymph, the potassium-rich fluid that bathes and nourishes the organ of Corti and provides a medium that conveys mechanical vibrations to the sensory hair cells. The cochlear duct plus the semicircular duct comprise the membranous labyrinth, a sac suspended in the bony canals of the inner ear. Endolymph is also present in the semicircular duct, but little is known of how it is formed there. Endolymph drains via the endolymphatic duct into the endolymphatic sac, which is lined by cells similar in structure to those of gallbladder. These cells probably absorb the endolymph (123, 170).

The stria vascularis derives its name from the unique association of transporting cells with vascular supply. Thin capillaries (cap) are embedded in the epithelium and are thus in intimate contact with the transporting cells. The stria is composed of two cell types. Marginal cells (Fig. 90, MC) have extensively folded lateral and basal plasma membranes (bf) dividing the cytoplasm into narrow compartments containing numerous

filamentous mitochondria (Fig. 92). The basement membranes of the capillary endothelial cells and of the marginal cells are in close contact and fuse in some regions (arrows, Fig. 91). The basal surface of the marginal cells is so highly folded that its basement membrane fills an extensive branching channel system, 0.3–0.6 $\mu$ wide, coursing through the epithelium (Fig. 92) (114, 232).

Basal cells (BC) contain few organelles and have processes extending toward the apical surface, separating the marginal cells (144, 232). In at least some species basal cells also face the lumen of the cochlear duct. The apical surface of the epithelium is relatively flat with occasional short microvilli.

Unlike most vertebrate epithelia, the stria vascularis uses potassium instead of sodium to generate a flow of water. The unique capillary arrangement may provide the cells with an ample supply of potassium (a) which is a minor component of plasma. In addition, potassium in the vascular tissue beneath the epithelium could be exchanged for sodium (b) across the basal cells. Potassium transport into the marginal cells (c) could bring about water entry into the cells. Flow across the apical surface may be passive (d) so the endolymph is simply an extension of the intracellular fluid of the stria cells (129).

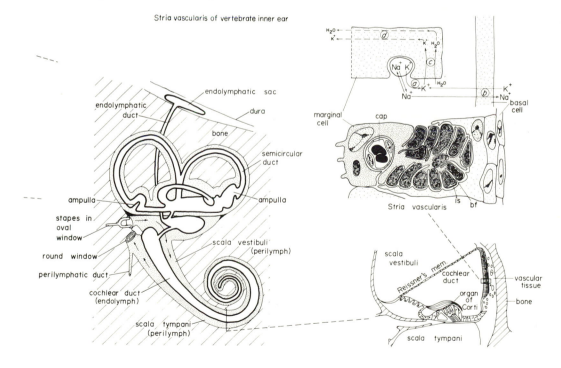

Fig. 90.  Stria vascularis of the cat inner ear, with the lumen or cochlear duct shown at top. ×3100. From *Amer. J. Anat.* **118**, 631–664. Courtesy of Dr. Raul Hinojosa.

Fig. 91.  The interface between the capillary endothelium and the stria vascularis cells in the guinea pig. ×26,000. Courtesy of Dr. Bernard Tandler.

Fig. 92.  Basal region of the stria vascularis of the cat. ×15,500. From *Amer. J. Anat.* **118**, 631–664. Courtesy of Dr. Raul Hinojosa.

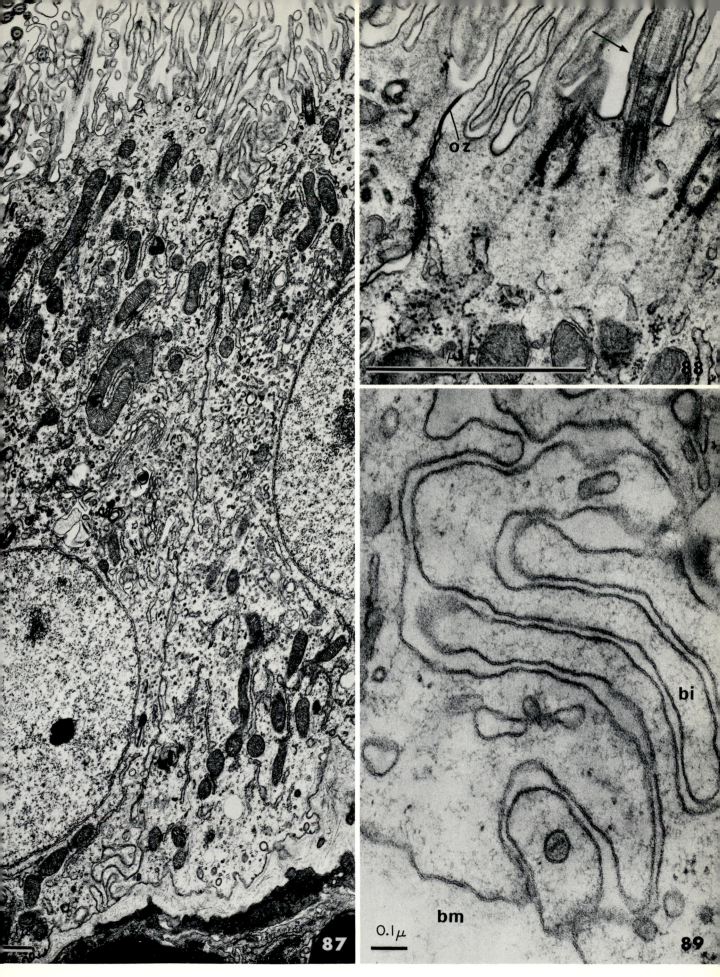

oz

88

bi

bm

0.1μ

87

89

## SPECIAL SECRETIONS

The evolution of complex sensory and nervous systems in the higher animals has been paralleled by development of a variety of accessory secretory epithelia. These produce the fluids that bathe and nourish the sophisticated sensory and integrating apparatus. The vertebrate choroid plexus secretes a fluid (cerebrospinal fluid) that provides the constant milieu for the brain and spinal neurons. Endolymph secreted by the stria vascularis plays an important role in sound perception by the organ of Corti of the vertebrate inner ear. The mechanism of endolymph secretion has not been established but we have suggested a mechanism based on the structure of this unusual epithelium. There are three main transporting epithelia associated with the function of the vertebrate eye. The ciliary epithelium secretes aqueous humor which bathes the inside of the eye, the corneal epithelium and mesothelium maintains the corneal stroma at the proper hydration for transparency, and the lacrimal glands secrete tears, which moisten and wash the outside of the eyeball. The lacrimal gland is not treated in detail as it is similar in structure to salivary glands, and little is known of the physiology of secretion.

The glands of the skin are also included in this section. Sweat glands and mammary glands resemble each other both in structure and mode of development. The fluid secreted by sweat glands plays an important role in thermoregulation, while mammary glands secrete a nutrient-rich fluid.

## Choroid Plexus

The choroid plexus is responsible for secreting the large volume of cerebrospinal fluid (CSF) which bathes the brain. CSF provides a constant ionic environment, removes waste products, and provides buoyancy and protection for the brain (219). In mammals, the choroid plexus secretes CSF into the ventricles. The fluid then flows into the cisterna magna and percolates over the surface of the brain in the subarachnoid space. CSF is finally returned to the bloodstream by a valve-like mechanism located within the arachnoid villi (299).

The four choroids are outpouchings of the pial vascular system lining the two lateral, third, and fourth ventricles. The general organization of the cells is similar in different vertebrates (180, 182, 237). The epithelial cells connect by occluding zonules (oz) in the apical region and also by ordinary desmosomes (d) at certain points along the lateral plasma membrane. The latter is usually straight but is somewhat folded in the basal region in some species. The basal plasma membrane is straight but may also be folded near the junction with the lateral membrane (Figs. 87, 89). The apical plasma membrane has numerous bulbous or clavate microvilli. Cilia are present in a few species (Fig. 88, arrow) (67, 191, 206, 231).

A characteristic feature of CSF is its constant composition despite wide fluctuations in blood composition. Although CSF resembles somewhat an ultrafiltrate of the blood, the fluid is probably produced by cell secretion rather than by pressure filtration. The structural basis of secretion by the choroid plexus has not been determined, but osmotic filtration (p. 5) may be involved. The absence of extensive basal infoldings would allow the cells to be more selective in the solutes they absorb from the blood.

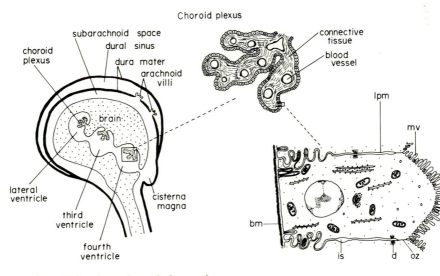

Fig. 87. Survey of the whole choroid epithelium of a chicken. ×14,000. Courtesy of Wesley J. Birge and Paul F. Doolin.

Fig. 88. Apical surface of the choroid plexus. ×58,000. Courtesy of Paul F. Doolin and Wesley J. Birge (*J. Cell Biol.* **29**, 333).

Fig. 89. Basal region of avian choroid plexus, showing interdigitations of the lateral plasma membrane. ×93,000. Courtesy of Wesley J. Birge and Paul F. Doolin (*Microsc. Cryst. Front* **14**, 458).

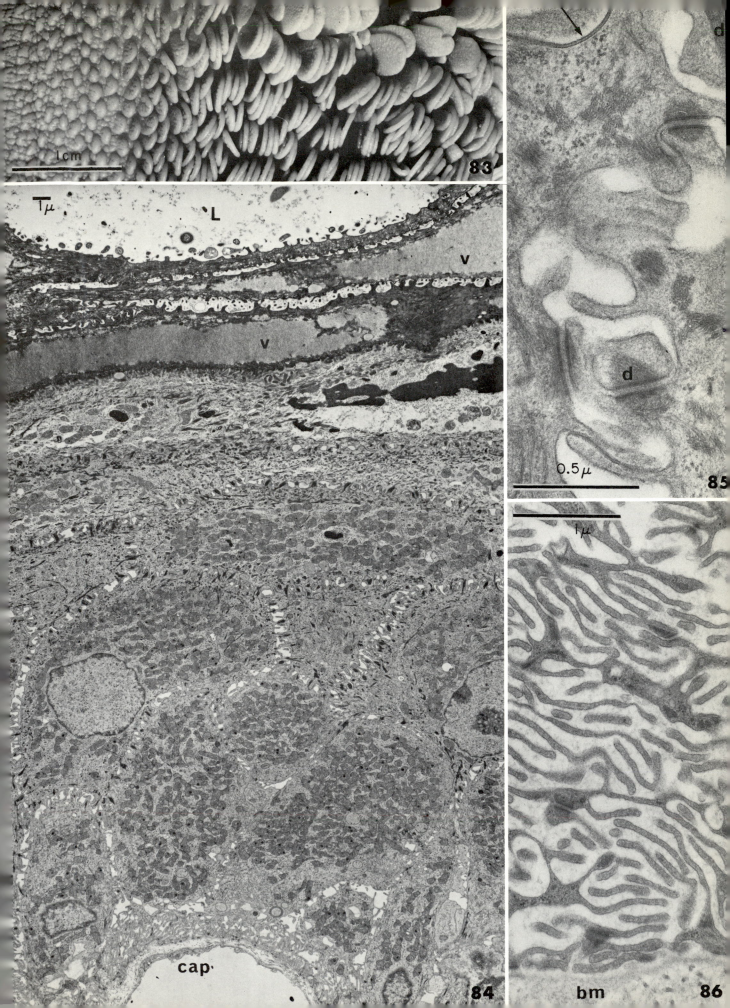

**83**

1 cm

**84**

Tμ L

v

v

cap

**85**

d

d

0.5 μ

**86**

1 μ

bm

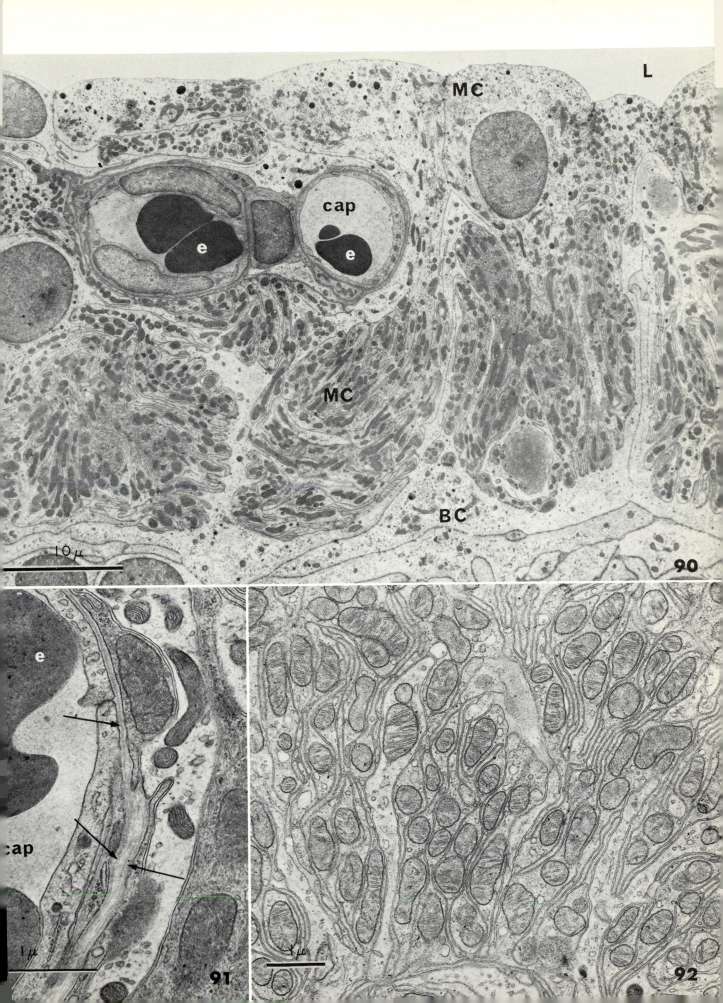

## Mammary Gland

Lactation in mammals is a complex phenomenon involving secretion of simple blood components (Na[+], K[+], Cl[-], Ca[2+], phosphate, citrate, and water) together with the synthesis and secretion of the specific milk constituents, lactose, casein, and fat (*164, 165*).

Milk is secreted by bulbous alveoli and flows into alveolar ducts which unite to form a single lactiferous duct opening at the surface. Alveoli are surrounded by stellate myoepithelial cells (Fig. 93, mc). Both lateral and basal membranes are infolded (Fig. 93) while the apical membrane has a few short microvilli (mv). The cell interior contains organelles associated with the formation and secretion of various milk components (*8, 9, 111, 115, 300*).

*Lipid* precursors entering the cell are synthesized (a) into lipid granules by the rough endoplasmic reticulum (rer) (*263*). The granules coalesce to form large granules (0.5–8 μ) which are pinched off (Fig. 94) from the apical surface (apocrine secretion). These lipid globules (lg) are bounded by a plasma membrane and a small amount of cytoplasm is often included (*154, 310*).

*Amino acids* are synthesized (b) into proteins by the rough endoplasmic reticulum (rer), packaged in the Golgi complex (G), and released to the lumen as protein granules 300–600 Å in diameter (Figs. 93 and 94, pg).

*Glucose* is absorbed from the blood and transformed into lactose by a lactose synthetase located on the Golgi complex membranes (c). Lactose may enter the lumen in the same vacuoles (v) as the protein granules. The secretion of lactose, which is the major osmotic component of milk, may provide most of the osmotic gradient for a passive flow of water. Since the ducts are impermeable to this disaccharide (*165*) the primary process of fluid formation may occur in the alveoli. Linzell and Peaker (*165*) suggest that as lactose enters the protein vacuoles, water follows by osmosis (d) and causes the vacuoles to swell. The water is then released into the lumen when the vacuoles fuse with the apical plasma membrane.

Mammary gland

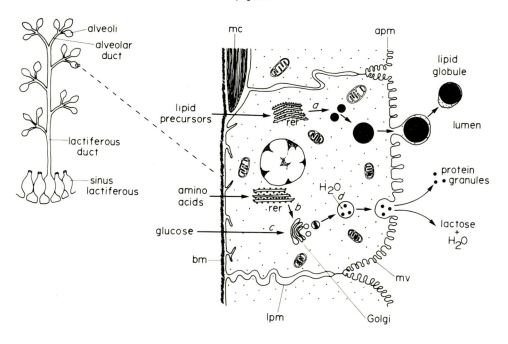

Fig. 93. The mammary gland of the guinea pig. ×11,000. Courtesy of Dr. F. B. P. Wooding.

Fig. 94. Secretion of lipid globules by the mammary gland of a goat. ×16,750. Courtesy of Dr. F. B. P. Wooding.

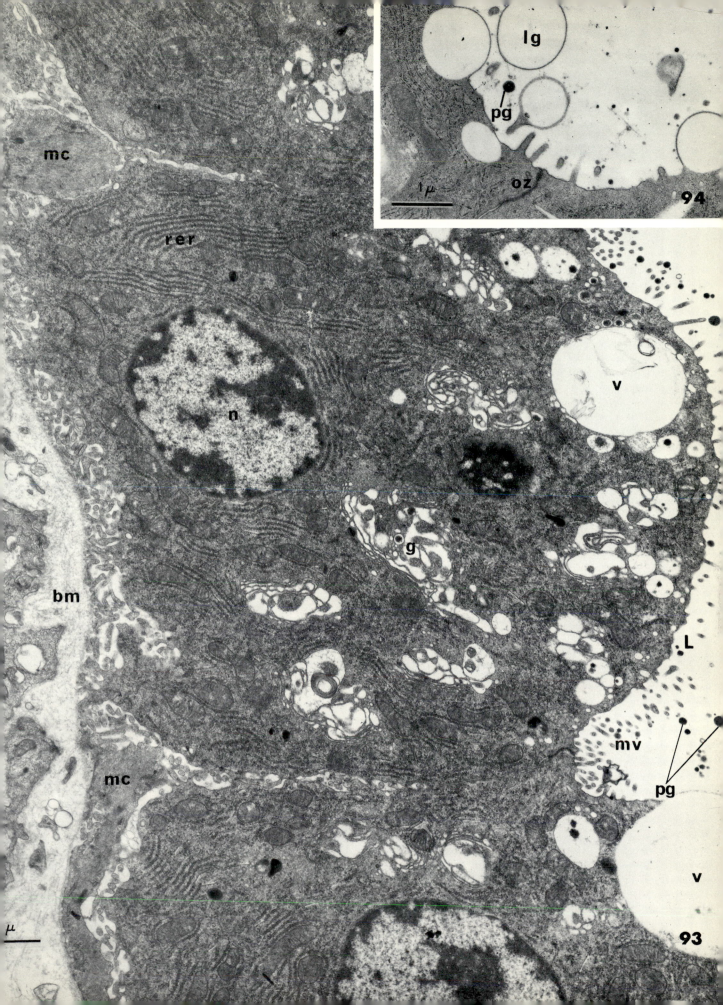

## Sweat Glands

Schiefferdecker (239) proposed two categories of sweat glands, eccrine and apocrine. Since then many types of glands have been described and there is much overlap between the two categories. Yet the terms are still used for convenience. Eccrine sweat is a watery fluid that functions in thermal regulation. Psychogenic sweat on palms and foot pads may aid in grip and improve tactile perception. Apocrine glands do not begin to function until puberty, and are generally regarded as scent glands. However, they are thermogenic in some species, as in the horse.

The sweat glands of man (illustrated here) are coiled tubular structures embedded in the dermis and opening onto the skin surface (eccrine) or into follicular canals (apocrine). The secretory region of both types of gland has basally located myoepithelial cells (mc) (72) and microvilli face the lumen. In eccrine glands the secretory region is composed of about equal numbers of serous cells and dark mucous cells. The serous cells have elaborately interdigitating basal and lateral membranes and intercellular canaliculi (∗) delimited by

junctional complexes (73, 74). The cytoplasm is rich in glycogen. Compared to the pale serous cells, mucous cells have relatively flat lateral and basal membranes. They secrete an epithelial mucin which contributes to the PAS positive material coating the inner surface of the sweat duct (75). In contrast, the secretory region of apocrine glands consists of a single type of cuboidal to columnar cell which has a highly folded basal surface (bi, Fig. 96) but rather straight lateral membranes. In both types of gland the ducts are double-layered and lack myoepithelial cells. They consist of basal cells and cuticular cells, the latter lining the lumen.

The clear serous cells with their elaborate canaliculi are responsible for producing the watery portion of eccrine sweat. The primary secretion is slightly hyperosmotic to blood and becomes hypoosmotic as it passes through the ducts, which reabsorb sodium chloride (248). Secretion of fluid in the apocrine glands probably involves equilibration across the basal infolds and apical microvilli. The apocrine mechanism, in which a portion of the cell detaches to form the secretion, may be an artifact (184, 185) or it may be valid for some sweat glands (146).

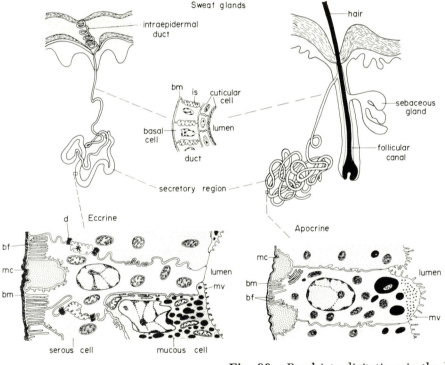

Fig. 95. Survey view of the human eccrine sweat gland, lumen at right and basal surface and myoepithelial cells at left. ×10,200. From Zelickson, ed. "Ultrastructure of Normal and Abnormal Skin," Lea & Febiger, Philadelphia. Courtesy of Dr. Richard A. Ellis.

Fig. 96. Basal interdigitations in the human apocrine sweat gland. ×22,640. From Zelickson, ed. "Ultrastructure of Normal and Abnormal Skin," Lea & Febiger, Philadelphia. Courtesy of Dr. Richard A. Ellis.

Fig. 97. Survey of the human apocrine sweat gland. ×8160. From Zelickson, ed. "Ultrastructure of Normal and Abnormal Skin," Lea & Febiger, Philadelphia. Courtesy of Dr. Richard A. Ellis.

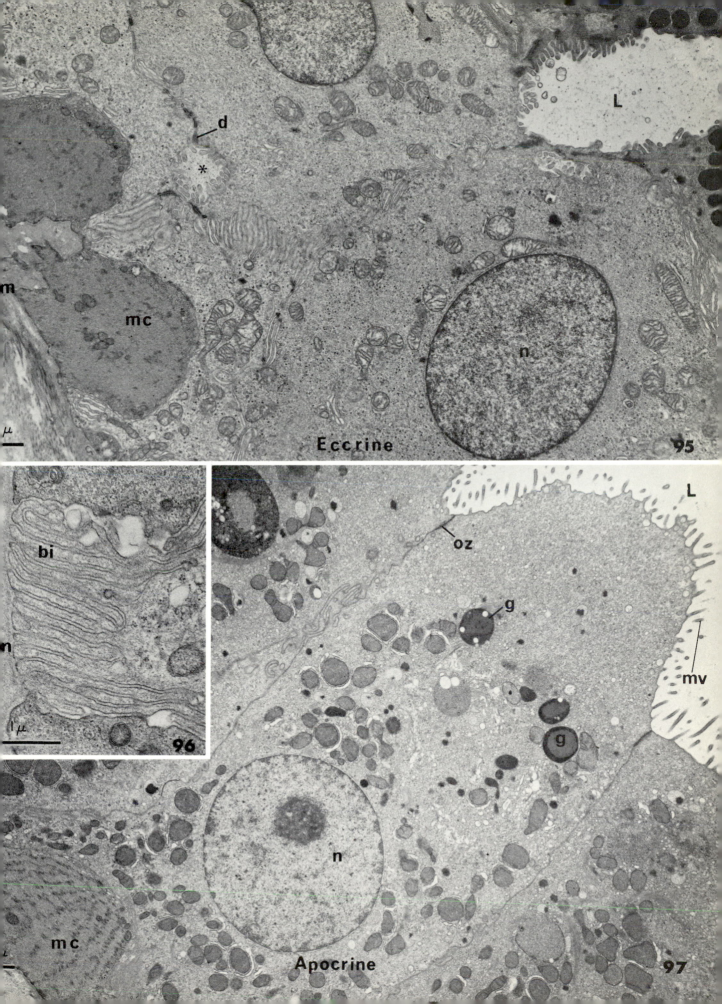

m

d

*

mc

m

mc

μ

n

L

Eccrine

95

bi

m

1 μ

96

oz

g

L

g

mv

n

mc

Apocrine

97

## Ciliary Epithelium of Vertebrate Eye

The ciliary epithelium of the eye secretes aqueous humor. This fluid is formed continuously and serves to maintain the intraocular volume and pressure within narrow limits while nourishing the avascular lens and cornea (54, 156). Sodium chloride, ascorbic acid, and certain amino acids enter the posterior chamber while iodide and certain organic anions are reabsorbed. Some of the aqueous humor diffuses away through the vitreous humor (a), but most of it flows between the ligaments of the lens (ciliary zonule), across the iris, and into the anterior chamber (b). From the anterior chamber, the aqueous humor drains into the canal of Schlemm (c), a thin-walled vein extending circumferentially all the way around the eye. From the canal of Schlemm the fluid enters the aqueous veins.

The ciliary epithelium (Fig. 98) consists of an inner pigmented layer (pe) and an outer nonpigmented layer (npe). The pigmented cells are generally cuboidal while the nonpigmented cells are more columnar, being about twice as high as wide. The two layers acquire an unusual relationship with one another during embryogenesis, so that the epithelium has two basement membranes (bm). The basal surface of the pigmented layer faces the highly vascular stroma, while the basal surface of the nonpigmented cells faces the posterior chamber. Apical surfaces of both layers are connected by desmosomes (d) and there is a ciliary canal system between them (Fig. 99) (7, 183, 275). The nature of this canal is unclear at present. The basal or secretory surface of the nonpigmented layer is highly folded (Fig. 100) and the cells interdigitate (bi) forming a complex labyrinth of long narrow channels opening in the direction of fluid flow (276). There is some folding of the basal surface of the pigmented cells as well.

The osmotic pressure difference between blood and the aqueous humor is insufficient to explain the observed rates of aqueous humor formation (6). Osmotic gradients may be confined to the labyrinthine channels (bi) of the nonpigmented layer. Ion transport into these channels would create local osmotic gradients for the passive flow of water (6).

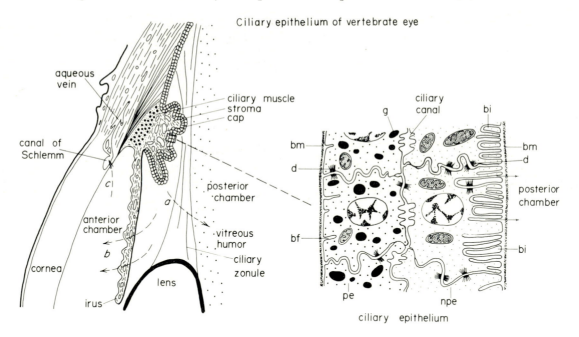

Ciliary epithelium of vertebrate eye

ciliary epithelium

**Fig. 98.** The ciliary epithelium of the adult albino rabbit, illustrating the pigmented layer on left and the nonpigmented layer on right. ×3700. From *Trans. Amer. Acad. Ophthalmol. Otolaryngol.* Sept.–Oct. 1966. Courtesy of Dr. John McD. Tormey.

**Fig. 99.** High magnification of the ciliary canal between the pigmented layer and nonpigmented layer. ×27,500. From *J. Cell Biol.* **17**, 641–659. Courtesy of Dr. John McD. Tormey.

**Fig. 100.** High magnification of the apical surface of the nonpigmented epithelium facing the posterior chamber of the eye. ×25,500. From *J. Cell Biol.* **23**, 658–664. Courtesy of Dr. John McD. Tormey.

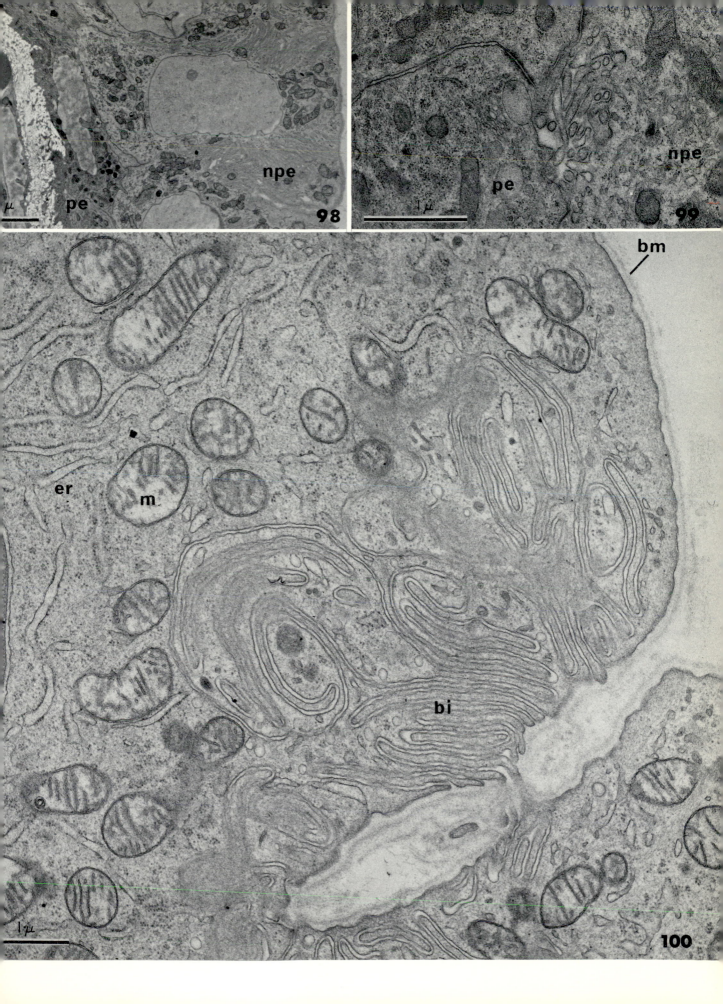

μ

pe

npe

**98**

1 μ

pe

npe

**99**

bm

er

m

bi

1 μ

**100**

## Cornea

The cornea is the transparent portion of the fibrous tunic covering the eyeball. Its outer surface is bathed with tear fluid secreted by the lacrimal gland, while its inner surface faces the anterior chamber, containing aqueous humor. The general structure of the cornea is similar in all vertebrates, being composed of an outer epithelium, a thick avascular stroma, and a single-layered "endothelium" or, more correctly, mesothelium, facing the aqueous humor (64).

### Epithelium

The stratified squamous epithelium (Fig. 102) is similar in over-all structure to frog skin, consisting of basal, intermediate, and flattened cornified cells (134). The tear surface (ts) is relatively smooth, with some short microvilli (Fig. 103) and occasional fine scales (20). The cells are connected by occluding zonules and desmosomes as in the frog skin. The epithelium is innervated by unmyelinated axons, has a thin basement membrane, and an underlying 6–9 $\mu$ thick layer of fibrillar material known as Bowman's membrane.

### Mesothelium

The corneal mesothelium (Fig. 101) is a single layer of flattened cells with a general resemblance to the peritoneal mesothelium (20, 124, 134). The cells contain an abundance of mitochondria and endoplasmic reticulum, the latter involved in the maintenance of Descemet's membrane (Des) (125), which is composed of a collagen-like protein. Adjacent cells are joined by occluding zonules near the surface facing the anterior chamber (ac). A vesicular transport system conveys tracers injected into the anterior chamber around the junctions and into the intercellular spaces from which they can diffuse into the stroma (134, 135).

The corneal stroma (st) must be maintained in a dehydrated state or it will swell and become nontransparent. Interest in corneal physiology centers about the possible roles of the two cellular layers in maintaining transparency. In rabbits, Na$^+$ is actively transported from the tear surface toward the aqueous humor side, the latter being maintained at a positive potential with respect to the former (65, 66). The potential is abolished by removing the epithelium but is unaffected by removing the endothelium (65). Calcium is also transported inward (80). In frogs, chloride is transported from the aqueous humor to the tear surface (312) while sodium may be transported in the opposite direction (33). Donn (64) proposes that Na$^+$ pumped into the stroma may neutralize excess negative charges on a cross-linked mucopolysaccharide. When the Na$^+$ influx is stopped, the negative charges are no longer neutralized, so they repel each other, the mucopolysaccharide uncoils, water enters, and the stroma swells.

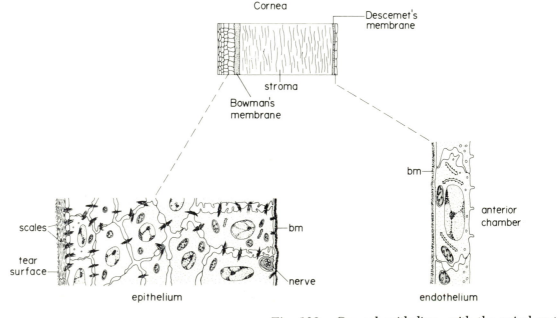

**Fig. 101.** The corneal mesothelium or "endothelium," Descemet's membrane, and stroma of adult human. ×14,800. From *Invest. Ophthalmol.* 43, 270–284. Courtesy of Drs. T. Iwamoto and George K. Smelser.

**Fig. 102.** Corneal epithelium, with the apical or tear surface at the top. ×5000. Courtesy of Drs. T. Iwamoto and George K. Smelser.

**Fig. 103.** High magnification of the corneal or tear surface, showing microvilli with fuzzy coat. ×105,300. Courtesy of Drs. T. Iwamoto and George K. Smelser.

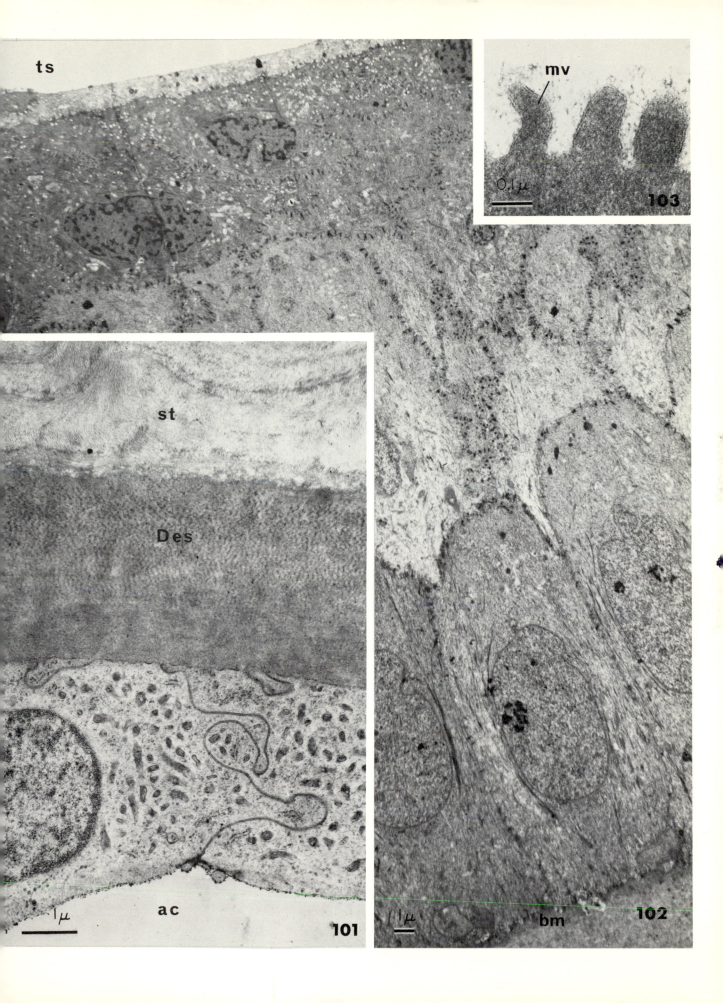

ts

mv

0.1μ

103

st

Des

1μ

ac

101

1μ

bm

102

# REFERENCES

1. Abel, J. H., and Ellis, R. A. (1966). Histochemical and electron microscopic observations on the salt secreting lacrymal glands of marine turtles. *Amer. J. Anat.* **118,** 337–358.

2. Altner, H. (1968). Die Ultrastruktur der Labialnephridien von *Onchiurus quadriocellatus* (Collembola). *J. Ultrastruct. Res.* **24,** 349–366.

3. Andersen, B., and Ussing, H. H. (1957). Solvent drag on non-electrolytes during osmotic flow through isolated toad skin and its response to antidiuretic hormone. *Acta Physiol. Scand.* **39,** 228–239.

4. Anderson, E., and Harvey, W. R. (1966). Active transport by the Cecropia midgut. II. Fine structure of the midgut epithelium. *J. Cell Biol.* **31,** 107–134.

5. Anderson, W. P., Aikman, D. P., and Meiri, A. (1970). Excised root exudation—a standing-gradient osmotic flow. *Proc. Roy. Soc. Ser. B.* **174,** 445–485.

6. Auricchio, G., and Bárány, E. H. (1959). On the role of osmotic water transport in the secretion of the aqueous humour. *Acta Physiol. Scand.* **45,** 190–210.

7. Bairati, A., and Orzalesi, N. (1966). The ultrastructure of the epithelium of the ciliary body: A study of the junctional complexes and of the changes associated with the production of plasmoid aqueous humor. *Z. Zellforsch. Mikrosk. Anat.* **69,** 636–658.

8. Bargmann, W., and Knoop, A. (1959). Über der Morphologie der Milchsekretion. Licht- und elektronenmikroskopische Studien an der Milchdrüse der Ratte. *Z. Zellforsch. Mikrosk. Anat.* **49,** 344–388.

9. Bargmann, W., Fleischhauer, K., and Knoop, A. (1961). Über die Morphologie der Milchsekretion. II. Zugleich eine Kritik am Schema der Sekretionsmorphologie. *Z. Zellforsch. Mikrosk. Anat.* **53,** 545–568.

10. Beams, H. W., Tahmisian, T. N., and Devine, R. L. (1955). Electron microscope studies on the cells. of the Malpighian tubules of the grasshopper (Orthoptera, Acrididae). *J. Biophys. Biochem. Cytol.* **1,** 197–202.

11. Bennett, H. S. (1963). Morphological aspects of extracellular polysaccharides. *J. Histochem. Cytochem.* **11,** 14–23.

12. Bentley, P. J. (1966). The physiology of the urinary bladder of amphibia. *Biol. Rev.* **41,** 275–316.

13. Berliner, R. W., Levinski, N. G., Davidson, D. G., and Eden, M. (1958). Dilution and concentration of the urine and the action of antidiuretic hormone. *Amer. J. Med.* **24,** 730–744.

14. Berridge, M. J. (1970). A structural analysis of intestinal absorption. *In* "Insect Ultrastructure" A. C. Neville, ed., pp. 135–151. Blackwell, Oxford.

15. Berridge, M. J. (1970). The role of 5-hydroxytryptamine and cyclic AMP in the control of fluid secretion by isolated salivary glands. *J. Exp. Biol.* **53,** 171–186.

16. Berridge, M. J. (1970). Unpublished results.

17. Berridge, M. J., and Gupta, B. L. (1967). Fine-structural changes in relation to ion and water transport in the rectal papillae of the blowfly, *Calliphora. J. Cell Sci.* **2,** 89–112.

18. Berridge, M. J., and Gupta, B. L. (1968). Fine-structural localization of adenosine triphosphatase in the rectum of *Calliphora. J. Cell Sci.* **3,** 17–32.

19. Berridge, M. J., and Oschman, J. L. (1969). A structural basis for fluid secretion by Malpighian tubules. *Tissue Cell* **1,** 247–272.

20. Blümcke, S., and Morgenroth, K. (1967). The stereo ultrastructure of the external and internal surface of the cornea. *J. Ultrastruct. Res.* **18,** 502–518.

21. Boler, R. K. (1969). Fine structure of canine Kupffer cells and their microtubule-containing cytosomes. *Anat. Rec.* **163,** 483–496.

22. Bonsdorff, C.-H. v., and Telkkä, A. (1966). The flagellar structure of the flame cell in fish tapeworm (*Diphyllobothrium latum*). *Z. Zellforsch. Mikrosk. Anat.* **70,** 169–179.

23. Brandenburg, J., and Kümmel, G. (1961). Die Feinstruktur der Solenocyten. *J. Ultrastruct. Res.* **5,** 437–452.

24. Braun, G., Kümmel, G., and Mangos, J. A. (1966). Studies on the ultrastructure and function of a primitive excretory organ, the protonephridium of the rotifer *Asplanchna priodonta. Pfluegers Arch. Gesamte Physiol.* **289,** 141–154.

25. Brightman, M. W., and Reese, T. S. (1969). Junctions between intimately apposed cell membranes in the vertebrate brain. *J. Cell Biol.* **40,** 648–677.

26. Bruns, R. R., and Palade, G. E. (1968). Studies on blood capillaries. I. General organization of blood capillaries in muscle. *J. Cell Biol.* **37,** 244–299.

27. Bulger, R. E. (1963). Fine structure of the rectal (salt-secreting) gland of the spiny dogfish, *Squalus acanthias. Anat. Rec.* **147,** 95–127.

28. Bulger, R. E. (1965). Electron microscopy of the stratified epithelium lining the excretory canal of the dogfish rectal gland. *Anat. Rec.* **151,** 589–607.

29. Bulger, R. E. (1965). The fine structure of the aglomerular nephron of the toadfish, *Opsanus tau. Amer. J. Anat.* **117,** 171–192.

30. Bulger, R. E. (1971). Personal communication.

31. Burger, J. W., and Hess, W. N. (1960). Function of the rectal gland in the spiny dogfish. *Science* **131,** 670–671.

32. Burkel, W. E., and Low, F. N. (1966). The fine structure of rat liver sinusoids, space of Disse and associated tissue space. *Amer. J. Anat.* **118,** 769–784.

33. Candia, O. A., and Askew, W. A. (1968). Active sodium transport in the isolated bullfrog cornea. *Biochim. Biophys. Acta* **163,** 262–265.

34. Carasso, N., Favard, P., and Valérien, J. (1962). Variations des ultrastructures dans les cellules épithéliales de la vessie du crapaud après stimulation par l'hormone neurohypophysaire. *J. Microsc. Paris* **1,** 143–158.

35. Carasso, N., Favard, P., Bourguet, J., and Jard, S. (1966). Role du flux net d'eau dans les modifications ultrastructurales de la vessie de grenouille stimulée par l'ocytocine. *J. Microsc. Paris* **5,** 519–522.

36. Cardell, R. R., Badenhausen, S., and Porter, K. R. (1967). Intestinal triglyceride absorption in the rat. An electron microscopical study. *J. Cell Biol.* **34,** 123–155.

37. Case, R. M., Harper, A. A., and Scratcherd, T. (1969). Water and electrolyte secretion by the pancreas. *In* "The Exocrine Glands" (Botelho S. Y. *et al.*, eds.), pp. 39–56. Univ. of Pennsylvania Press, Philadelphia.

38. Chalcroft, J. P., and Bullivant, S. (1970). An interpretation of liver cell membrane and junction structure based on observation of freeze-fracture replicas of both sides of the fracture. *J. Cell Biol.* **47,** 49–60.

39. Chan, D. K. O., and Phillips, J. G. (1967). The anatomy, histology and histochemistry of the rectal gland in the lip-shark *Hemiscyllium plagiosum* (Bennett). *J.Anat.* **101,** 137–157.

40. Chapman, G. B., Chiarodo, A. J., Coffey, R. J., and Wieneke, K. (1966). The fine structure of mucosal epithelial cells of a human pathological gall bladder. *Anat. Rec.* **154,** 579–615.

41. Choi, J. K. (1963). The fine structure of the urinary bladder of the toad, *Bufo marinus. J. Cell Biol.* **16,** 53–72.

42. Clement, P. (1968). Ultrastructures d'un rotifère: *Notammata copeus*. I. La cellule-flamme. Hypothèses physiologiques. *Z. Zellforsch. Mikrosk. Anat.* **89,** 478–498.

43. Cope, F. W. (1969). Ion and water transport across multicellular membranes through extracellular space by chemiperistaltic waves. *Bull. Math. Biophys.* **31,** 529–540.

44. Copeland, D. E. (1948). The cytological basis of chloride transfer in the gills of *Fundulus heteroclitus. J. Morphol.* **82,** 201–227.

45. Copeland, D. E. (1964). A mitochondrial pump in the cells of the anal papillae of mosquito larvae. *J. Cell. Biol.* **23,** 253–263.

46. Copeland, D. E., and Fitzjarrell, A. T. (1968). The salt absorbing cells in the gills of the blue crab (*Callinectes sapidus* Rathbun) with notes on modified mitochondria. *Z. Zellforsch. Mikrosk. Anat.* **92,** 1–22.

47. Crane, R. K. (1968). Absorption of sugars. *In* "Handbook of Physiology" (C. F. Code, ed.), Sect. 6, Vol. III. pp. 1323–1351. Amer. Physiol. Soc., Washington.

48. Crofts, D. R. (1925). The comparative morphology of the caecal gland (rectal gland) of selachian fishes, with some reference to the morphology and physiology of the similar intestinal appendage throughout Ichthyopsida and Sauropsida. *Proc. Zool. Soc. London* **Part I,** 101–188.

49. Curran, P. F. (1960). Na, Cl, and water transport by rat ileum *in vitro. J. Gen. Physiol.* **43,** 1137–1148.

50. Curran, P. F. (1965). Ion transport in intestine and its coupling to other transport processes. *Fed. Proc. Fed. Amer. Soc. Exp. Biol.* **24,** 993–999.

51. Curran, P. F., and MacIntosh, J. R. (1962). A model system for biological water transport. *Nature (London)* **193,** 347–348.

52. Darnton, S. J. (1969). A possible correlation between ultrastructure and function in the thin descending and ascending limbs of the loop of Henle of rabbit kidney. *Z. Zellforsch. Mikrosk. Anat.* **93,** 516–524.

53. Davis, L. E., and Schmidt-Nielsen, B. (1967). Ultrastructure of the crocodile kidney (*Crocodylus acutus*) with special reference to electrolyte and fluid transport. *J. Morphol.* **121,** 255–276.

54. Davson, H. (1956). "Physiology of the Ocular and Cerebrospinal Fluids." Little, Brown, Boston, Massachusetts.

55. Diamond, J. M. (1968). Transport mechanisms in the gallbladder. *In* "Handbook of Physiology" (C. F. Code, ed.), Sec. 6 Vol. V. pp. 2451–2482. Amer. Physiol. Soc., Washington.

56. Diamond, J. M., and Bossert, W. H. (1967). Standing-gradient osmotic flow. A mechanism for coupling of water and solute transport in epithelia. *J. Gen. Physiol.* **50,** 2061–2083.

57. Diamond, J. M., and Bossert, W. H. (1968). Functional consequences of ultrastructural geometry in "backwards" fluid-transporting epithelia. *J. Cell Biol.* **37,** 694–702.

58. Diamond, J. M., and Tormey, J. McD. (1966). Role of long extracellular channels in fluid transport across epithelia. *Nature (London)* **210,** 817–820.

59. Diamond, J. M., and Tormey, J. McD. (1966). Studies on the structural basis of water transport across epithelial membranes. *Fed. Proc. Fed. Amer. Soc. Exp. Biol.* **25,** 1458–1463.

60. Diamond, J. M., Barry, P. H., and Wright, E. M. (1971). The route of transepithelial ion permeation in the gallbladder. *In* "Electrophysiology of Epithelial Cells" (G. Giebisch, ed.), pp. 23–33. Schattauer, Stuttgart.

61. DiBona, D. R., Civan, M. M., and Leaf, A. (1969). The anatomic site of the transepithelial permeability barriers of toad bladder. *J. Cell Biol.* **40,** 1–7.

62. Dietschy, J. M. (1966). Recent developments in solute and water transport across the gall bladder epithelium. *Gastroenterology* **50,** 692–707.

63. Dobson, A., and Phillipson, A. T. (1968). Absorption from the ruminant forestomach. *In* "Handbook of Physiology," Vol. V, Sect. 6, Chap. 132, pp. 2761–2774. Amer. Physiol. Soc., Washington.

64. Donn, A. (1966). Cornea and sclera (review). *Arch. Ophthalmol.* **75,** 261–288.

65. Donn, A., Maurice, D. M., and Mills, N. L. (1959). Studies on the living cornea *in vitro*. I. Method and physiologic measurements. *Arch. Ophthalmol.* **62,** 741–747.

66. Donn, A., Maurice, D. M., and Mills, N. L. (1959). Studies on the living cornea *in vitro*. II. The active transport of sodium across the epithelium. *Arch. Ophthalmol.* **62,** 748–757.

67. Doolin, P. F., and Birge, W. J. (1969). Ultrastructural organization and histochemical profile of adult fowl choroid plexus epithelium. *Anat. Rec.* **165,** 515–530.

68. Dougherty, R. W. (1965). *In* "Physiology of Digestion in the Ruminant" (R. W. Dougherty, ed.). Butterworths, Washington.

69. Doyle, W. L. (1960). The principal cells of the salt-gland of marine birds. *Exp. Cell Res.* **21,** 386–393.

70. Doyle, W. L. (1962). Tubule cells of the rectal salt-gland of *Urolophus. Amer. J. Anat.* **111,** 223–237.

71. Ekholm, R., and Edlund, Y. (1959). Ultrastructure of the human exocrine pancreas. *J. Ultrastruct. Res.* **2,** 453–481.

72. Ellis, R. A. (1965). Fine structure of the myoepithelium of the eccrine sweat glands of man. *J. Cell Biol.* **27,** 551–563.

73. Ellis, R. A. (1967). Eccrine, sebaceous and apocrine glands. In "Ultrastructure of Normal and Abnormal Skin" (A. S. Zelickson, ed.). Lea & Febiger, Philadelphia, Pennsylvania.

74. Ellis, R. A. (1968). Eccrine sweat glands: Electron microscopy, cytochemistry and anatomy. *Handb. d. Haut-u. Geschlechtskr.* 1, 224–266.

75. Ellis, R. A., and Montagna, W. (1961). Electron microscopy of the duct, and especially the "cuticular border" of the eccrine sweat glands in *Macaca mulatta*. *J. Biophys. Biochem. Cytol.* 9, 238–242.

76. Ericsson, J. L. E., and Olsen, S. (1970). On the fine structure of the aglomerular renal tubule in *Lophius piscatorius*. *Z. Zellforsch. Mikrosk. Anat.* 104, 240–258.

77. Ericsson, J. L. E., and Trump, B. F. (1964). Electron microscopic study of the epithelium of the proximal tubule of the rat kidney. I. The intracellular localization of acid phosphatase. *Lab. Invest.* 13, 1427–1456.

78. Ernst, S. A., and Ellis, R. A. (1969). The development of surface specialization in the secretory epithelium of the avian salt gland in response to osmotic stress. *J. Cell Biol.* 40, 305–321.

79. Evett, R. D., Higgins, J. A., and Brown, A. L. (1964). The fine structure of normal mucosa in human gall bladder. *Gastroenterology* 47, 49–60.

80. Ezuka, K., Kikkawa, Y., Kurosawa, K., and Okada, N. (1960). Active transport of Ca ions across the rabbit cornea. *Jap. J. Physiol.* 10, 204–210.

81. Fänge, R., Schmidt-Nielsen, K., and Osaki, H. (1958). The salt gland of the herring gull. *Biol. Bull.* 115, 162–171.

82. Fänge, R., Schmidt-Nielsen, K., and Robinson, M. (1958). Control of secretion from the avian salt gland. *Amer. J. Physiol.* 195, 321–326.

83. Farquhar, M. G., and Palade, G. E. (1963). Junctional complexes in various epithelia. *J. Cell Biol.* 17, 375–412.

84. Farquhar, M. G., and Palade, G. E. (1964). Functional organization of amphibian skin. *Proc. Nat. Acad. Sci. U.S.* 51, 569–577.

85. Farquhar, M. G., and Palade, G. E. (1965). Cell junctions in amphibian skin. *J. Cell. Biol.* 26, 263–291.

86. Farquhar, M. G., and Palade, G. E. (1966). Adenosine triphosphatase localization in amphibian epidermis. *J. Cell Biol.* 30, 359–379.

87. Farquhar, M. G., Wissig, S. L., and Palade, G. E. (1961). Glomerular permeability. I. Ferritin transfer across the normal glomerular capillary wall. *J. Exp. Med.* 113, 47–66.

88. Ferguson, D. R., and Heap, P. F. (1970). The morphology of the toad urinary bladder: A stereoscopic and transmission electron microscopical study. *Z. Zellforsch. Mikrosk. Anat.* 109, 297–305.

89. Fordtran, J. S., and Dietschy, J. M. (1966). Water and electrolyte movement in the intestine. *Gastroenterology* 50, 263–285.

90. Frederiksen, O., and Leyssac, P. P. (1969). Transcellular transport of isosmotic volumes by the rabbit galbladder *in vitro*. *J. Physiol.* 201, 201–224.

91. Frömter, E., and Diamond, J. M. (1972). Route of passive ion permeation in epithelia. *Nature (London)* 235, 9–13.

92. Giebisch, G. (1969). Functional organization of proximal and distal tubular electrolyte transport. *Nephron* 6, 260–281.

93. Gilula, N. B., Branton, D., and Satir, P. (1970). The septate junction: a structural basis for intercellular coupling. *Proc. Nat. Acad. Sci. U. S.* 67, 213–220.

94. Goodenough, D. A., and Revel, J. P. (1970). A fine structural analysis of intercellular junctions in the mouse liver. *J. Cell Biol.* 45, 272–290.

95. Gouranton, J. (1967). Élaboration d'une mucoprotéine acide dans l'appareil de Golgi des cellules d'une portion de l'intestin moyen de divers Cercopidae. *C. R. Acad. Sci. Ser. D* 264, 2584–2587.

96. Grantham, J. J., Ganote, C. E., Burg, M. B., and Orloff, J. (1969). Paths of trans-tubular water flow in isolated renal collecting tubules. *J. Cell Biol.* 41, 562–576.

97. Grantham, J. J., Cuppage, F. E., and Fanestil, D. (1971). Direct observation of toad bladder response to vasopressin. *J. Cell Biol.* 48, 695–699.

98. Grimstone, A. V., Mullinger, A. M., and Ramsay, J. A. (1968). Further studies on the rectal complex of the mealworm *Tenebrio molitor*, L. (Coleoptera, Tenebrionidae). *Phil. Trans. Roy. Soc. B.* 253, 343–382.

99. Groepler, W. (1969). Feinstruktur der Coxalorgane bei der Gattung *Ornithodorus* (Acari: Argasidae). *Z. Wiss. Zool. Abt. A*, 178, 235–275.

100. Gupta, B. L., and Berridge, M. J. (1966). A coat of repeating subunits on the cytoplasmic surface of the plasma membrane in the rectal papillae of the blowfly, *Calliphora erythrocephala*. (Meig.), studied in situ by electron microscopy. *J. Cell Biol.* 29, 376–382.

101. Gupta, B. L., and Berridge, M. J. (1966). Fine structural organization of the rectum in the blowfly, *Calliphora erythrocephala* (Meig) with special reference to connective tissue, tracheae and neurosecretory innervation in the rectal papillae. *J. Morphol.* 120, 23–82.

102. Harvey, W. R., Haskell, J. A., and Nedergaard, S. (1968). Active transport by the Cecropia midgut. III. Midgut potential generated directly by active K-transport. *J. Exp. Biol.* 48, 1–12.

103. Harvey, W. R., and Nedergaard, S. (1964). Sodium-independent active transport of potassium in the isolated midgut of the Cecropia silkworm. *Proc. Nat. Acad. Sci. U.S.* 51, 757–765.

104. Haskell, J. A., Clemons, R. D., and Harvey, W. R. (1965). Active transport by the Cecropia midgut. I. Inhibitors, stimulants and potassium-transport. *J. Cell. Comp. Physiol.* 65, 45–56.

105. Haupt, J. (1969). Zur Feinstruktur der Maxillarnephridien von *Scutigerella immaculata* Newport (Symphyla, Myriapoda). *Z. Zellforsch. Mikrosk. Anat.* 101, 401–407.

106. Haupt, J. (1969). Zur Feinstruktur der Labialniere des Silberfischchens *Lepisma saccharina* L. (Thysanura, Insecta). *Zool. Beitr. N.F.* 15, 139–170.

107. Hayward, A. F. (1962). Aspects of the fine structure of the gall bladder epithelium of the mouse. *J. Anat.* **96**, 227–236.

108. Heath, T., and Wissig, S. L. (1966). Fine structure of the surface of mouse hepatic cells. *Amer. J. Anat.* **119**, 97–128.

109. Hecker, H., Diehl, P. A., and Aeschlimann, A. (1969). Recherches sur l'ultrastructure et l'histochimie de l'organe coxal d'*Ornithodorus moubata* (Murray) (Ixodoidea; Argasidae). *Acta Trop.* **26**, 346–360.

110. Helander, H. F. (1962). Ultrastructure of fundus glands of the mouse gastric mucosa. *J. Ultrastruct. Res. Suppl.* **4**, 1–123.

111. Helminen, H. J., and Ericsson, J. L. E. (1968). Studies on mammary gland involution. I. On the ultrastructure of the lactating mammary gland. *J. Ultrastruct. Res.* **25**, 193–213.

112. Henrikson, R. C. (1970). Ultrastructure of ovine ruminal epithelium and localization of sodium in the tissue. *J. Ultrastruct. Res.* **30**, 385–401.

113. Henrikson, R. C., and Stacy, B. D. (1971). The barrier to diffusion across ruminal epithelium: A study by electron microscopy using horseradish peroxidase, lanthanum, and ferritin. *J. Ultrastruct. Res.* **34**, 72–82.

114. Hinojosa, R., and Rodriguez-Echandia, E. L. (1966). The fine structure of the stria vascularis of the cat inner ear. *Amer. J. Anat.* **118**, 631–664.

115. Hollman, K. H. (1959). L'ultrastructure de la glande mammaire normale de la souris en lactation. Etude au microscope électronique. *J. Ultrastruct. Res.* **2**, 423–443.

116. Hollman, K. H., and Verley, J. M. (1965). La glande sous-maxillaire de la souris et du rat. Étude au microscope électronique. *Z. Zellforsch. Mikrosk. Anat.* **68**, 363–388.

117. Hughes, G. M., and Grimstone, A. V. (1965). The fine structure of the secondary lamellae of the gills of *Gadus pollachius*. *Quart. J. Microsc. Sci.* **106**, 343–353.

118. Hungate, R. E. (1968). Ruminal fermentation. *In* "Handbook of Physiology," Vol. V, Sect. 6, Ch. 130, pp. 2725–2745. *Amer. Physiol. Soc.*, Washington.

119. Ito, S. (1970). Personal communication.

120. Ito, S. (1961). The endoplasmic reticulum of gastric parietal cells. *J. Cell Biol.* **11**, 333–347.

121. Ito, S. (1965). The enteric surface coat on cat intestinal microvilli. *J. Cell. Biol.* **27**, 475–491.

122. Ito, S., and Winchester, R. J. (1963). The fine structure of the gastric mucosa in the bat. *J. Cell Biol.* **16**, 541–577.

123. Iurato, S. (1967). "Submicroscopic Structure of the Inner Ear." Pergamon, Oxford.

124. Iwamoto, T., and Smelser, G. K. (1965). Electron microscopy of the human corneal endothelium with reference to transport mechanisms. *Invest. Ophthalmol.* **43**, 270–284.

125. Jacus, M. A. (1956). Studies on the cornea: II. The fine structure of the Descemet's membrane. *J. Biophys. Biochem. Cytol.* **2** (suppl.), 243–255.

126. Janowitz, H. D. (1967). Pancreatic secretion of fluid and electrolytes. *In* "Handbook of Physiology," Ch. 52, Sect. 6, Vol. II. pp. 925–933. Amer. Physiol. Soc., Washington.

127. Johnson, F. R., and Darnton, S. J. (1967). Ultrastructural observations on the renal papilla of the rabbit. *Z. Zellforsch. Mikrosk. Anat.* **81**, 390–406.

128. Johnson, F. R., McMinn, R. M. H., and Birchenough, R. F. (1962). The ultrastructure of the gall-bladder epithelium of the dog. *J. Anat.* **96**, 447–487.

129. Johnstone, B. M. (1964). Endolymph and endocochlear potentials. *In* "Transcellular Membrane Potentials and Ionic Fluxes" (F. M. Snell and W. K. Noell, eds.). Gordon & Breach, New York.

130. Jones, A. L., and Fawcett, D. W. (1966). Hypertrophy of the agranular endoplasmic reticulum in hamster liver induced by phenobarbital (with a review on the function of this organelle in liver). *J. Histochem. Cytochem.* **14**, 215–232.

131. Jones, J. C., and Zeve, V. H. (1968). The fine structure of the gastric caeca of *Aedes aegypti* larvae. *J. Insect Physiol.* **14**, 1567–1575.

132. Kaye, G. I. (1962). Studies on the cornea. III. The fine structure of the frog cornea and the uptake and transport of colloidal particles by the cornea *in vitro*. *J. Cell Biol.* **15**, 241–258.

133. Kaye, G. I., and Lane, N. (1965). The epithelial basal complex: A morphophysiological unit in transport and absorption. *J. Cell Biol.* **27**, 50A.

134. Kaye, G. I., and Pappas, G. D. (1962). Studies on the cornea. I. The fine structure of the rabbit cornea and the uptake and transport of colloidal particles by the cornea *in vivo*. *J. Cell Biol.* **12**, 457–479.

135. Kaye, G. I., Pappas, G. D., Donn, A., and Mallett, N. (1962). Studies on the cornea. II. The uptake and transport of colloidal particles by the living rabbit cornea *in vitro*. *J. Cell Biol.* **12**, 481–501.

136. Kaye, G. I., Wheeler, H. O., Whitlock, R. T., and Lane, N. (1966). Fluid transport in the rabbit gallbladder. A combined physiological and electron microscopic study. *J. Cell Biol.* **30**, 237–268.

137. Kelly, D. E. (1966). Fine structure of desmosomes, hemidesmosomes, and an adepidermal globular layer in developing newt epidermis. *J. Cell Biol.* **28**, 51–72.

138. Kendall, M. D. (1969). The fine structure of the salivary glands of the desert locust, *Schistocerca gregaria* Forskål. *Z. Zellforsch. Mikrosk. Anat.* **98**, 399–420.

139. Kern, H. F. and Ferner, H. (1971). Die Feinstruktur des exokrinen Pankreasgewebes vom Menschen. *Z. Zellforsch. Mikrosk. Anat.* **113**, 322–343.

140. Kessel, R. G., and Beams, H. W. (1962). Electron microscope studies on the gill filaments of *Fundulus heteroclitus* from sea water and fresh water with special reference to the ultrastructural organization of the "chloride cell." *J. Ultrastruct. Res.* **6**, 77–87.

141. Kessel, R. G., and Beams, H. W. (1963). Electron microscope observations on the salivary gland of the cockroach, *Periplaneta americana*. *Z. Zellforsch. Mikrosk. Anat.* **59**, 857–877.

142. Keynes, R. D. (1969). From frog skin to sheep rumen: A survey of transport of salts and water across multicellular structures. *Quart. Rev. Biophys.* **2**, 177–281.

143. Keys, A., and Willmer, E. N. (1932). "Chloride secreting cells" in the gills of fishes, with special reference to the common eel. *J. Physiol.* **76**, 368–378.

144. Kikuchi, K., and Hilding, D. A. (1966). The development of the stria vascularis in the mouse. *Acta Otolaryngol.* **62**, 277–291.

145. Kirschner, L. B. (1967). Comparative physiology: Invertebrate excretory organs. *Annu. Rev. Physiol.* **29**, 169–196.

146. Kneeland, J. E. (1966). Fine structure of the sweat glands of the antebrachial organ of *Lemur catta*. *Z. Zellforsch. Mikrosk. Anat.* **73**, 521–533.

147. Koefoed-Johnsen, V., and Ussing, H. H. (1958). The nature of the frog skin potential. *Acta Physiol. Scand.* **42**, 298–308.

148. Komnick, H. (1963). Elektronenmikroskopische Untersuchungen zur funktionellen Morphologie des Ionentransportes in der Salzdrüse von *Larus argentatus*. III. Teil: Funktionelle Morphologie der Tubulusepithelzellen. *Protoplasma* **56**, 605–636.

149. Komnick, H., and Wohlfarth-Bottermann, K. E. (1966). Zur Cytologie der Rectaldrüsen von Knorpelfischen. I. Teil: Die Feinstruktur der Tubulusepithelzellen. *Z. Zellforsch. Mikrosk. Anat.* **74**, 123–144.

150. Krogh, A. (1937). Osmotic regulation in fresh water fishes by active absorption of chloride ions. *Z. Vergl. Physiol.* **24**, 656–666.

151. Kuhn, N. O., and Olivier, M. L. (1965). Ultrastructure of the hepatic sinusoid of the goat *Capra hircus*. *J. Cell Biol.* **26**, 977–979.

152. Kümmel, G. (1958). Das Terminalorgan der Protonephridien, Feinstruktur und Deutung der Funktion. *Z. Naturforsch.* **13b**, 677–679.

153. Kümmel, G. (1964). Das Cölomsachen der Antennendrüse von *Cambarus affinis* Say. (Decapoda, Crustacea). *Zool. Beitr.* **10**, 227–252.

154. Kurosumi, K., Kobayashi, Y., and Baba, N. (1968). The fine structure of the mammary glands of lactating rats, with special reference to the apocrine secretion. *Exp. Cell Res.* **50**, 177–192.

155. Kurtz, S. M. (1964). The salivary glands. *In* "Electron Microscopic Anatomy" (S. M. Kurtz, ed.). Academic Press, New York.

156. Langham, M. E. (1958). Aqueous humor and control of intra-ocular pressure. *Physiol. Rev.* **38**, 215–242.

157. Latta, H., Maunsbach, A. B., and Madden, S. C. (1960). The centrolobular region of the renal glomerulus studied by electron microscopy. *J. Ultrastruct. Res.* **4**, 455–472.

158. Latta, H., Maunsbach, A. B., and Osvaldo, L. (1967). The fine structure of renal tubules in cortex and medulla. *In* "Ultrastructure of the Kidney," (A. J. Dalton and F. Haguenau, eds.), pp. 1–56. Academic Press, New York.

159. Lees, A. D. (1946). Chloride regulation and the function of the coxal glands in ticks. *Parasitology* **37**, 172–184.

160. van Lennep, E. W. (1968). Electron microscopic histochemical studies on salt-excreting glands in elasmobranchs and marine catfish. *J. Ultrastruct. Res.* **25**, 94–108.

161. van Lennep, E. W., and Komnick, H. (1970). Fine structure of the nasal salt gland in the desert lizard *Uromastyx acanthinurus*. *Cytobiologie* **1**, 47–67.

162. Lillibridge, C. B. (1964). The fine structure of normal human gastric mucosa. *Gastroenterology* **47**, 269–290.

163. Lillibridge, C. B. (1968). Electron microscopic measurements of the thickness of various membranes in oxyntic cells from frog stomachs. *J. Ultrastruct. Res.* **23**, 243–259.

164. Linzell, J. L. (1959). Physiology of the mammary glands. *Physiol. Rev.* **39**, 534–576.

165. Linzell, J. L., and Peaker, M. (1971). Mechanism of milk secretion. *Physiol. Rev.* **51**, 564–597.

166. Locke, M. (1965). The structure of septate desmosomes. *J. Cell Biol.* **25**, 166–169.

167. Lockwood, A. P. M. (1968). Aspects of the Physiology of Crustacea. Oliver & Boyd, Edinburgh and London.

168. Loewenstein, W. R. (1966). Permeability of membrane junctions. *Ann. N. Y. Acad. Sci.* **137**, 441–472.

169. Loewenstein, W. R., and Kanno, Y. (1964). Studies on an epithelial (gland) cell junction. I. Modifications of surface membrane permeability. *J. Cell Biol.* **22**, 565–586.

170. Lundquist, P.-G. (1965). The endolymphatic duct and sac in the guinea pig. *Acta Oto-laryngol. Suppl.* **201**, 1–108.

171. McCuskey, R. S. (1966). A dynamic and static study of hepatic arterioles and hepatic sphincters. *Amer. J. Anat.* **119**, 455–478.

172. McKanna, J. A. (1968). Fine structure of the protonephridial system in Planaria. I. Flame cells. *Z. Zellforsch. Mikrosk. Anat.* **92**, 509–523.

173. McNutt, N. S., and Weinstein, R. S. (1970). The ultrastructure of the nexus. A correlated thin-section and freeze-cleave study. *J. Cell Biol.* **47**, 666–688.

174. Machen, T. E., and Diamond, J. M. (1969). An estimate of the salt concentration in the lateral intercellular spaces of rabbit gallbladder during maximal fluid transport. *J. Membrane Biol.* **1**, 194–213.

175. Maetz, J. (1968). Salt and water metabolism. *In* "Perspectives in Endocrinology" (E. J. W. Barrington and C. B. Jørgensen, eds.), pp. 47–162. Academic Press, New York.

176. Mangos, J. A., Braun, G., and Hamann, K. F. (1966). Micropuncture study of sodium and potassium excretion in the rat parotid saliva. *Pfluegers Arch.* **291**, 99–106.

177. Marshall, E. K., and Grafflin, A. L. (1928). The structure and function of the kidney of *Lophius piscatorius*. *Bull. Johns Hopkins Hosp.* **43**, 205–235.

178. Matter, A., Orci, L., and Rouiller, C. (1969). A study on the permeability barriers between Disse's space and the bile canaliculus. *J. Ultrastruct. Res. Suppl.* **11**, 1–71.

179. Maunsbach, A. B. (1969). Functions of lysosomes in kidney cells. *In* "Lysosomes in Biology and Pathology," Vol. I, pp. 115–154. North-Holland Publ., Amsterdam.

180. Maxwell, D. S., and Pease, D. C. (1956). The electron microscopy of the choroid plexus. *J. Biophys. Biochem. Cytol.* **2**, 467–474.

181. Menefee, M. G., and Meuller, C. B. (1967). Some morphological considerations of transport in the glomerulus. *In* "Ultrastructure of the Kidney" (A. J. Dalton and F. Haguenau, eds.), pp. 73–100. Academic Press, New York.

182. Millen, J. W., and Rogers, G. E. (1956). An electron microscopic study of the choroid plexus in the rabbit. *J. Biophys. Biochem. Cytol.* **2**, 407–416.

183. Missotten, L. (1964). L'Ultrastructure de tissus oculaires. *Bull. Soc. Belge Ophthalmol.* **136**, 1–200.

184. Munger, B. L. (1965). The cytology of apocrine sweat glands. I. Cat and monkey. *Z. Zellforsch. Mikrosk. Anat.* **67**, 373–389.

185. Munger, B. L. (1965). The cytology of apocrine sweat glands. II. Human. *Z. Zellforsch. Mikrosk. Anat.* **68**, 837–851.

186. Nakao, T. (1965). The excretory organ of *Amphioxus* (*Branchiostoma*) *belcheri*. *J. Ultrastruct. Res.* **12**, 1–12.

187. Newstead, J. D. (1967). Fine structure of the respiratory lamellae of teleostean gills. *Z. Zellforsch. Mikrosk. Anat.* **79**, 396–428.

188. Noirot, C., Noirot-Timothée, C., and Kovoor, J. (1967). Revêtement particulaire de la membrane plasmatique en rapport avec l'excrétion dans une région spécialisée de l'intestin moyen des Termites supérieurs. *C. R. Acad. Sci.* **264**, 722–725.

189. Noirot-Timothée, C., and Noirot, C. (1965). L'Intestin moyen chez la reine des termites supérieurs. *Ann. Sci. Naturelles, Zool. Paris 12ᵉ Série VII*, 185–208.

190. Ogilvie, J. T., McIntosh, J. R., and Curran, P. F. (1963). Volume flow in a series-membrane system. *Biochim. Biophys. Acta* **66**, 441–444.

191. Oksche, A., and Vaupel-von Harnack, M. (1969). Elektronenmikroskopische Studien über Altersveränderungen (Filamente) der Plexus chorioidei des Menschen (Biopsiematerial). *Z. Zellforsch. Mikrosk. Anat.* **93**, 1–29.

192. Olsen, S. (1966). Ultrastructure of the base of renal tubule cells of marine teleosts. *Nature (London)* **212**, 95–96.

193. Olsen, S., and Ericsson, J. L. E. (1968). Ultrastructure of the tubule of the aglomerular teleost *Nerophis ophidion*. *Z. Zellforsch. Mikrosk. Anat.* **87**, 17–30.

194. O'Riordan, A. M. (1969). Electrolyte movement in the isolated midgut of the cockroach (*Periplaneta americana* L.) *J. Exp. Biol.* **51**, 699–714.

195. Orrenius, S., and Ericsson, J. L. E. (1966). Enzyme-membrane relationship in phenobarbital induction of synthesis of drug-metabolizing enzyme system and proliferation of endoplasmic membranes. *J. Cell Biol.* **28**, 181–198.

196. Oschman, J. L. (1972). The cell surface of an insect intestine, (in preparation).

197. Oschman, J. L., and Berridge, M. J. (1970). Structural and functional aspects of salivary fluid secretion in *Calliphora*. *Tissue Cell* **2**, 281–310.

198. Oschman, J. L., and Berridge, M. J. (1971). The structural basis of fluid secretion. *Fed. Proc. Fed. Amer. Soc. Exp. Biol.* **30**, 49–56.

199. Oschman, J. L., and Wall, B. J. (1969). The structure of the rectal pads of *Periplaneta americana* L. with regard to fluid transport. *J. Morphol.* **127**, 475–509.

200. Osvaldo, L., and Latta, H. (1966). The thin limbs of the loop of Henle. *J. Ultrastruct. Res.* **15**, 144–168.

201. Pak Poy, R. F. K., and Bentley, P. J. (1960). Fine structure of the epithelial cells of the toad urinary bladder. *Exp. Cell Res.* **20**, 235–237.

202. Palay, S. L., and Karlin, L. J. (1959). An electron microscopic study of the intestinal villus. I. The fasting animal. *J. Biophys. Biochem. Cytol.* **5**, 363–372.

203. Palay, S. L., and Karlin, L. J. (1959). An electron microscopic study of the intestinal villus. II. The pathway of fat absorption. *J. Biophys. Biochem. Cytol.* **5**, 373–384.

204. Parisi, M., Ripoche, P., Bourguet, J., Carasso, N., and Favard, P. (1969). The isolated epithelium of frog urinary bladder. Ultrastructural modifications under the action of oxytocin, theophylline and cyclic AMP. *J. Microsc. Paris* **8**, 1031–1036.

205. Parks, H. F. (1961). On the fine structure of the parotid gland of mouse and rat. *Amer. J. Anat.* **108**, 303–329.

206. Paul, E. (1968). Histochemische Studien an den Plexus chorioidei, an der Paraphyse und am Ependym von *Rana temporaria* L. *Z. Zellforsch. Mikrosk. Anat.* **91**, 519–546.

207. Peachey, L. D., and Rasmussen, H. (1961). Structure of the toad's urinary bladder as related to its physiology. *J. Biophys. Biochem. Cytol.* **10**, 529–553.

208. Peaker, M. (1971). Intracellular concentrations of sodium, potassium and chloride in the salt-gland of the domestic goose and their relation to the secretory mechanism. *J. Physiol.* **213**, 399–410.

209. Pease, D. C. (1968). Myoid features of renal corpuscles and tubules. *J. Ultrastruct. Res.* **23**, 304–320.

210. Phillips, J. E. (1964). Rectal absorption in the desert locust, *Schistocerca gregaria* Forskål. I. Water. *J. Exp. Biol.* **41**, 15–38.

211. Phillips, J. E., and Dockrill, A. A. (1968). Molecular sieving of hydrophilic molecules by the rectal intima of the desert locust (*Schistocerca gregaria*). *J. Exp. Biol.* **48**, 521–532.

212. Philpott, C. W. (1965). Halide localization in the teleost chloride cell and its identification by selected area electron diffraction. *Protoplasma* **60**, 7–23.

213. Philpott, C. W., and Copeland, D. E. (1963). Fine structure of chloride cells from three species of *Fundulus*. *J. Cell Biol.* **18**, 389–404.

214. Pitts, R. F. (1968). "Physiology of the Kidney and Body Fluids." Year Book Medical Publ., Chicago, Illinois.

215. Pontin, R. M. (1964). A comparative account of the protonephridia of *Asplanchna* (Rotifera) with special reference to the flame bulbs. *Proc. Zool. Soc. London* **142**, 511–525.

216. Pontin, R. M. (1966). The osmoregulatory function of the vibratile flames and the contractile vesicle of *Asplanchna* (Rotifera). *Comp. Biochem. Physiol.* **17**, 1111–1126.

217. Porter, K. R. (1969). Independence of fat absorption and pinocytosis. *Fed. Proc. Fed. Amer. Soc. Exp. Biol.* **28**, 35–40.

218. Potts, W. T. W., and Parry, G. (1964). "Osmotic and Ionic Regulation in Animals." Pergamon, London.

219. Rall, D. P. (1967). Comparative pharmacology and cerebrospinal fluid. *Fed. Proc. Fed. Amer. Soc. Exp. Biol.* 26, 1020–1023.

220. Ramsay, J. A. (1950). Osmotic regulation in mosquito larvae. *J. Exp. Biol.* 27, 145–157.

221. Ramsay, J. A. (1953). Active transport of potassium by the Malpighian tubules of insects. *J. Exp. Biol.* 30, 358–369.

222. Ramsay, J. A. (1955). The excretory system of the stick insect, *Dixippus morosus* (Orthoptera, Phasmidae). *J. Exp. Biol.* 32, 183–199.

223. Ramsay, J. A. (1955). The excretion of sodium, potassium and water by the Malpighian tubules of the stick insect, *Dixippus morosus* (Orthoptera, Phasmidae) *J. Exp. Biol.* 32, 200–216.

224. Ramsay, J. A. (1964). The rectal complex of the mealworm *Tenebrio molitor*, L. (Coleoptera, Tenebrionidae). *Phil. Trans. Roy. Soc. London Ser. B* 248, 279–314.

225. Rasmont, R. (1960). Structure et ultrastructure de la glande coxale d'un scorpion. *Ann. Soc. Zool. Belge* 89, 239–268.

226. Revel, J. P., and Karnovsky, M. J. (1967). Hexagonal array of subunits in intercellular junctions of the mouse heart and liver. *J. Cell Biol.* 33, C7–C12.

227. Rhodin, J. (1958). Anatomy of kidney tubules. *Int. Rev. Cytol.* 7, 485–534.

228. Riegel, J. A. (1970). A new model of transepithelial fluid movement with detailed application to fluid movement in the crayfish antennal gland. *Comp. Biochem. Physiol.* 36, 403–410.

229. Ritch, R., and Philpott, C. W. (1969). Repeating particles associated with an electrolyte-transport membrane. *Exp. Cell Res.* 55, 17–24.

230. Robertson, J. D. (1963). The occurrence of a subunit pattern in the unit membranes of club endings in Mauthner cell synapses in goldfish brains. *J. Cell Biol.* 19, 201–221.

231. Rodriguez, E. M. (1967). Light and electron microscopy of granules in the toad choroid plexus. *Z. Zellforsch. Mikrosk. Anat.* 82, 362–375.

232. Rodriguez Echandia, E. L., and Burgos, M. H. (1965). The fine structure of the stria vascularis of the guinea pig inner ear. *Z. Zellforsch. Mikrosk. Anat.* 67, 600–619.

233. Rosa, F. (1963). Ultrastructure of the parietal cell of the human gastric mucosa in the resting state and after stimulation with histalog. *Gastroenterology* 45, 354–363.

234. Ross, M. H., and Reith, E. J. (1970). Myoid elements in the mammalian nephron and their relationship to other specializations in the basal part of kidney tubule cells. *Amer. J. Anat.* 129, 399–416.

235. Roth, T. F., and Porter, K. R. (1964). Yolk protein uptake in the oocyte of the mosquito *Aedes aegypti* L. *J. Cell Biol.* 20, 313–332.

236. Rubin, W. (1969). Enzyme cytochemistry of gastric parietal cells at a fine structure level. Cytochemical separation of the endoplasmic reticulum from the "tubulovesicles." *J. Cell Biol.* 42, 332–338.

237. Santolaya, R. C., and Rodriguez Echandia, E. L. (1968). The surface of the choroid plexus cell under normal and experimental conditions. *Z. Zellforsch. Mikrosk. Anat.* 92, 43–51.

238. Satir, P., and Gilula, N. B. (1970). The cell junction in a lamellibranch gill ciliated epithelium. *J. Cell Biol.* 47, 468–487.

239. Schiefferdecker, P. (1917). Die Hautdrüsen des Menschen und der Säugetiere, ihre biologische und rassenanatomische Bedeutung sowie de Muscularis sexualis. *Biol. Zentralbl.* 37, 534–562.

240. Schmidt-Nielsen, B., and Davis, L. E. (1968). Fluid transport and tubular intercellular spaces in reptilian kidneys. *Science* 159, 1105–1108.

241. Schmidt-Nielsen, B., Gertz, K. H., and Davis, L. E. (1968). Excretion and ultrastructure of the antennal gland of the fiddler crab *Uca mordax*. *J. Morphol.* 125, 473–495.

242. Schmidt-Nielsen, B., and Skadhauge, E. (1967). Function of the excretory system of the crocodile (*Crocodylus acutus*). *Amer. J. Physiol.* 212, 973–980.

243. Schmidt-Nielsen, K. (1963). Osmotic regulation in higher vertebrates. *Harvey Lect.* 58, 53–93.

244. Schmidt-Nielsen, K. (1965). Physiology of salt glands. *In* "Sekretion und Exkretion," pp. 269–282. Springer-Verlag, Berlin.

245. Schmidt-Nielsen, K., Jörgensen, C. B., and Osaki, H. (1958). Extrarenal salt excretion in birds. *Amer. J. Physiol.* 193, 101–107.

246. Schnorr, B., and Vollmerhaus, B. (1967). Die Feinstruktur des Pansenepithels von Ziege und Rind. *Zentralbl. Veterinaermed.* 14, 789–818.

247. Schultz, S. G., and Curran, P. F. (1968). Intestinal absorption of sodium chloride and water. *In* "Handbook of Physiology." Sect. 6, Vol. III, pp. 1245–1275. Amer. Physiol. Soc., Washington.

248. Schulz, I., Ullrich, K. J., Frömter, E., Holzgreve, H., Frick, A., and Hegel, U. (1965). Mikropunktion und elektrische Potentialmessung an Schweissdrüsen des Menschen. *Pfluegers Arch. Gesamte Physiol.* 284, 360–372.

249. Sedar, A. W. (1961). Electron microscopy of the oxyntic cell in the gastric glands of the bullfrog (*Rana catesbiana*). I. The non-acid-secreting gastric mucosa. *J. Biophys. Biochem. Cytol.* 9, 1–18.

250. Sedar, A. W. (1961). Electron microscopy of the oxyntic cell in the gastric glands of the bullfrog, *Rana catesbiana*. II. The acid-secreting gastric mucosa. *J. Cell Biol.* 10, 47–57.

251. Sedar, A. W. (1962). Electron microscopy of the oxyntic cell in the gastric glands of the bullfrog, *Rana catesbiana*. III. Permanganate fixation of the endoplasmic reticulum. *J. Cell Biol.* 14, 152–156.

252. Sedar, A. W. (1965). Fine structure of the stimulated oxyntic cell. *Fed. Proc. Fed. Amer. Soc. Exp. Biol.* 24, 1360–1367.

253. Sedar, A. W. (1969). Uptake of peroxidase into the smooth-surfaced tubular system of the gastric acid-secreting cell. *J. Cell Biol.* 43, 179–184.

254. Sedar, A. W. (1969). Electron microscopic demonstration of polysaccharides associated with acid-secreting cells of the stomach after "inert dehydration." *J. Ultra-*

*struct. Res.* **28**, 112–124.

255. Sedar, A. W., and Friedman, M. H. F. (1961). Correlation of the fine structure of the gastric parietal cell (dog) with functional activity of the stomach. *J. Cell Biol.* **11**, 349–363.

256. Silverblatt, F. J., and Bulger, R. E. (1970). Gap junctions occur in vertebrate renal proximal tubule cells. *J. Cell Biol.* **47**, 513–515.

257. Slautterback, D. B. (1965). Mitochondria in cardiac muscle cells of the canary and some other birds. *J. Cell Biol.* **24**, 1–21.

258. Smith, D. S. (1963). The structure of flight muscle sarcosomes in the blowfly *Calliphora erythrocephala* (Diptera) *J. Cell Biol.* **19**, 115–138.

259. Smith, D. S., and Littau, V. C. (1960). Cellular specialization in the excretory epithelia of an insect, *Macrosteles fascifrons* Stål (Homoptera). *J. Biophys. Biochem Cytol.* **8**, 103–133.

260. Smith, D. S., Compher, K., Janners, M., Lipton, C., and Wittle, L. W. (1969). Cellular organization and ferritin uptake in the mid-gut epithelium of a moth, *Ephestia kühniella. J. Morphol.* **127**, 41–72.

261. Smith, H. W. (1951). "The Kidney: Structure and Function in Health and Disease." Oxford Univ. Press, London and New York.

262. Stäubli, W., Freyvogel, T. A., and Suter, J. (1966). Structural modification of the endoplasmic reticulum of midgut epithelial cells of mosquitoes in relation to blood intake. *J. Microsc. Paris* **5**, 189–204.

263. Stein, O., and Stein, Y. (1967). Lipid synthesis, intracellular transport, and secretion. II. Electron microscopic radioautographic study of the mouse lactating mammary gland. *J. Cell Biol.* **34**, 251–263.

264. Stephens R. J. and Pfeiffer, C. J. (1968). Ultrastructure of the gastric mucosa of normal laboratory ferrets. *J. Ultrastruct. Res.* **22**, 45–62.

265. Stobbart, R. H. (1967). The effect of some anions and cations upon the fluxes and net uptake of chloride in the larvae of *Aëdes aegypti* (L.), and the nature of the uptake mechanisms for sodium and chloride. *J. Exp. Biol.* **47**, 35–57.

266. Straus, L. P. (1963). A study of the fine structure of the so-called chloride cell in the gill of the guppy *Lebistes reticulatus* P. *Physiol. Zool.* **36**, 183–198.

267. Strauss, E. W. (1968). Morphological aspects of triglyceride absorption. *In* "Handbook of Physiology." Sect. 6, Vol. III pp. 1377–1406. Amer. Physiol. Soc., Wash.

268. Tamarin, A. (1966). Myoepithelium of the rat submaxillary gland. *J. Ultrastruct. Res.* **16**, 320–338.

269. Tamarin, A., and Sreebny, L. M. (1966). The rat submaxillary salivary gland. A correlative study by light and electron microscopy. *J. Morphol.* **117**, 295–352.

270. Tandler, B. (1962). Ultrastructure of the human submaxillary gland. I. Architecture and histological relationships of the secretory cells. *Amer. J. Anat.* **111**, 287–307.

271. Tandler, B. (1963). Ultrastructure of the human submaxillary gland. II. The base of the striated duct cells. *J. Ultrastruct. Res.* **9**, 65–75.

272. Threadgold, L. T., and Houston, A. H. (1964). An electron microscope study of the "chloride cell" of *Salmo salar* L. *Exp. Cell Res.* **34**, 1–23.

273. Tisher, C. C., Bulger, R. E., and Trump, B. F. (1966). Human renal ultrastructure. I. Proximal tubule of healthy individuals. *Lab. Invest.* **15**, 1357–1394.

274. Tisher, C. C., Bulger, R. E., and Valtin, H. (1971). Morphology of renal medulla in water diuresis and vasopressin-induced antidiuresis. *Amer. J. Physiol.* **220**, 87–94.

275. Tormey, J. McD. (1963). Fine structure of the ciliary epithelium of the rabbit, with particular reference to "infolded membranes," "vesicles," and the effects of Diamox. *J. Cell Biol.* **17**, 641–659.

276. Tormey, J. McD. (1964). Differences in membrane configuration between osmium-tetroxide-fixed and glutaraldehyde-fixed ciliary epithelium. *J. Cell Biol.* **23**, 658–664.

277. Tormey, J. McD., and Diamond, J. M. (1967). The ultrastructural route of fluid transport in rabbit gallbladder. *J. Gen. Physiol.* **50**, 2031–2060.

278. Treherne, J. E. (1962). The physiology of absorption from the alimentary canal in insect. *In* "Viewpoints in Biology" (J. D. Carthy and C. L. Duddington, eds.), Vol. 1, pp. 201–241. Butterworths, London.

279. Treherne, J. E. (1965). Active transport in insects. *In* "Aspects of Insect Biochemistry" (T. W. Goodwin, ed.), pp. 1–13. Academic Press, New York.

280. Treherne, J. E. (1967). Gut absorption. *Annu. Rev. Entomol.* **12**, 43–58.

281. Trier, J. S. (1963). Studies on small intestinal crypt epithelium. I. The fine structure of the crypt epithelium of the proximal small intestine of fasting humans. *J. Cell Biol.* **18**, 599–620.

282. Trier, J. S. (1968). Morphology of the epithelium of the small intestine. *In* "Handbook of Physiology" Sect. 6, Vol. III, pp. 1125–1175. Amer. Physcol. Soc., Washington.

283. Trier, J. S., and Rubin, C. E. (1965). Electron microscopy of the small intestine: A review. *Gastroenterology* **49**, 574–603.

284. Turbeck, B. O., and Foder, B. (1970). Studies on a carbonic anhydrase from the midgut epithelium of larvae of Lepidoptera. *Biochim. Biophys. Acta* **212**, 139–149.

285. Tyson, G. E. (1968). The fine structure of the maxillary gland of the brine shrimp, *Artemia salina*: the end-sac. *Z. Zellforsch. Mikrosk. Anat.* **86**, 129–138.

286. Tyson, G. E. (1969). The fine structure of the maxillary gland of the brine shrimp, *Artemia salina*: the efferent duct. *Z. Zell. Mikrosk. Anat.* **93**, 151–163.

287. Ullrich, K. J., Kramer, K., and Boylan, J. W. (1961). Present knowledge of the counter-current system in the mammalian kidney. *Progr. Cardiov. Dis.* **3**, 395–431.

288. Ussing, H. H. (1970). The effect of urea on permeability and transport of frog skin. *In* "Urea and the Kidney" (B. Schmidt-Nielsen, ed.), *Excerpta Med. Int. Congr. Ser.* **195**, 138–148.

289. Ussing, H. H. (1970). Tracer studies and membrane structure. *In* "Capillary Permeability" (C. Crone and N. A. Lassen, eds.), pp. 654–656. Academic Press, New York.

290. Ussing, H. H., and Windhager, E. E. (1964). Nature of shunt path and active sodium transport path through frog skin epithelium. *Acta Physiol. Scand.* **61**, 484–504.

291. Vial, J. D., and Orrego, H. (1960). Electron microscope observations on the fine structure of parietal cells. *J. Biophys. Biochem. Cytol.* **7**, 367–372.

292. Voûte, C. L. (1963). An electron microscopic study of the skin of the frog (*Rana pipiens*). *J. Ultrastruct. Res.* **9**, 497–510.

293. Voûte, C. L., and Ussing, H. H. (1968). Some morphological aspects of active sodium transport. The epithelium of the frog skin. *J. Cell Biol.* **36**, 625–638.

294. Voûte, C. L., and Ussing, H. H. (1970). The morphological aspects of shunt-path in the epithelium of the frog skin (*R. temporaria*). *Exp. Cell Res.* **61**, 133–140.

295. Wall, B. J. (1971). Local osmotic gradients in the rectal pads of an insect. *Fed. Proc. Fed. Amer. Soc. Exp. Biol.* **30**, 42–48.

296. Wall, B. J., and Oschman, J. L. (1970). Water and solute uptake by rectal pads of *Periplaneta americana*. *Amer. J. Physiol.* **218**, 1208–1215.

297. Wall, B. J., Oschman, J. L., and Schmidt-Nielsen, B. (1970). Fluid transport: concentration of the intercellular compartment. *Science* **167**, 1497–1498.

298. Warner, F. D. (1969). The fine structure of the protonephridia in the rotifer *Asplanchna*. *J. Ultrastruct. Res.* **29**, 499–524.

299. Welch, K., and Friedman, V. (1960). The cerebrospinal fluid valves. *Brain* **83**, 454–469.

300. Wellings, S. R., and Philp, J. R. (1964). The function of the golgi apparatus in lactating cells of the BALB/c Crgl mouse. An electron microscopic and autoradiographic study. *Z. Zellforsch. Mikrosk. Anat.* **61**, 871–882.

301. Wigglesworth, V. B. (1932). On the function of the so-called 'rectal glands' of insects. *Quart. J. Microsc. Sci.* **75**, 131–150.

302. Wigglesworth, V. B. (1933). The function of the anal gills of the mosquito larva. *J. Exp. Biol.* **10**, 16–26.

303. Wigglesworth, V. B., and Salpeter, M. M. (1962). Histology of the Malpighian tubules in *Rhodnius prolixus* Stål (Hemiptera). *J. Insect Physiol.* **8**, 299–307.

304. Wilborn, W. H., and Shackleford, J. M. (1969). The cytology of submandibular gland of the opossum. *J. Morphol.* **128**, 1–34.

305. Wirz, H. (1957). The location of antidiuretic action in the mammalian kidney. *In* "The Neurohypophysis" (H. Heller, ed.), pp. 157–169. Butterworth, London.

306. Wood, J. L. (1971). Personal communication.

307. Wood, J. L., Farrand, P. S., and Harvey, W. R. (1969). Active transport of potassium by the Cecropia midgut. VI. Microelectrode potential profile. *J. Exp. Biol.* **50**, 169–178.

308. Wood, R. L. (1959). Intercellular attachment in the epithelium of *Hydra* as revealed by electron microscopy. *J. Biophys. Biochem. Cytol.* **6**, 343–352.

309. Wood, R. L. (1963). Evidence of species differences in the ultrastructure of the hepatic sinusoid. *Z. Zellforsch. Mikrosk. Anat.* **58**, 679–692.

310. Wooding, F. B. P., Peaker, M., and Linzell, J. L. (1970). Theories of milk secretion: Evidence from the electron microscopic examination of milk. *Nature (London)* **226**, 762–764.

311. Young, J. A., and Schögel, E. (1966). Micropuncture investigation of sodium and potassium excretion in rat submaxillary saliva. *Pfluegers Arch.* **291**, 85–98.

312. Zadunaisky, J. A. (1966). Active transport of chloride in frog cornea. *Amer. J. Physiol.* **211**, 506–512.

## SUPPLEMENTARY REFERENCES

### Review Articles

Botelho, S. Y., Brooks, F. P., and Shelley, W. B., eds. (1969). "Exocrine Glands." Univ. of Pennsylvania Press, Philadelphia.

Comar, C. L., and Bronner, F., eds. (1960). "Mineral Metabolism: An Advanced Treatise," 2 vols. Academic Press, New York.

Keynes, R. D. (1969). From frog skin to sheep rumen: A survey of transport of salts and water across multicellular structures. *Quart. Rev. Biophys.* **2**, 177–281.

Kirschner, L. B. (1967). Comparative physiology: Invertebrate excretory organs. *Annu. Rev. Physiol.* **29**, 169–196.

Krogh, A. (1939). "Osmotic Regulation in Aquatic Animals." Cambridge Univ. Press, London and New York.

Lefevre, P. G. (1955). Active transport through animal cell membranes. *Protoplasmatologia* **8** (7a).

Möllendorff, W. Von., ed. (1929). "Handbuch der mikroskopischen Anatomie des Menschen." J. Springer, Berlin.

Potts, W. T. W., and Parry, G. (1964). "Osmotic and Ionic Regulation in Animals." Pergamon, London.

Ramsay, J. A. (1961). The comparative physiology of renal function in invertebrates. *In* "The Cell and the Organism" (J. A. Ramsay and V. B. Wigglesworth, eds.), pp. 158–174. Cambridge Univ. Press, London and New York.

Ussing, H. H., Kruhøffer, P., Thaysen, J. H., and Thorn, N. A. (1960). The alkali metals in biology. *In* "Handbuch des experimentellen Pharmakologie," Vol. XIII. Springer, Berlin.

Wessing, A. (1968). Funktionsmorphologie von Exkretionsorganen bie Insekten. *Zool. Anz. Suppl.* **31**, 633–681.

### Protonephridia

Clément, P. (1969). Ultrastructures d'un Rotifère *Notommata copeus*. II. Le tube protonéphridien. *Z. Zellforsch. Mikrosk. Anat.* **94**, 103–117.

Kümmel, G. (1964). Die Feinstruktur der Terminalzellen (Cyrtocyten) an den Protonephridien der Priapuliden. *Z. Zellforsch. Mikrosk. Anat.* **62**, 468–484.

Kümmel, G., and Brandenburg, J. (1961). Die Reusengeiselzellen (Cyrtocyten). *Z. Naturforsch.* **16b**, 692–697.

McKanna, J. A. (1968). Fine structure of the protonephridial system in Planaria. II. Ductules, collecting ducts, and osmoregulatory cells. *Z. Zellforsch. Mikrosk. Anat.* **92**, 524–535.

Pedersen, K. J. (1961). Some observations on the fine structure of planarian protonephridia and gastrodermal phagocytes. *Z. Zellforsch. Mikrosk. Anat.* **53**, 609–628.

Race, G. J., Larsh, J. E., Esch, G. W., and Martin, J. H. (1965). A study of the larval stage of *Multiceps serialis* by electron microscopy. *J. Parasitol.* **51**, 364–369.

Senft, A. W., Philpott, D. E., and Pelofsky, A. H. (1961). Electron microscope observations of the integument, flame cells, and gut of *Schistosoma mansoni. J. Parasitol.* **47**, 217–229.

### Invertebrate Extrarenal Organs

Copeland, D. E. (1968). Fine structure of salt and water uptake in the land-crab, *Gecarcinus lateralis. Amer. Zoologist* **8**, 417–432.

Phillips, J. E., and Meredith, J. (1969). Active sodium and chloride transport by anal papillae of a salt water mosquito larva (*Aedes campestris*). *Nature (London)* **222**, 168–169.

Sohal, R. S., and Copeland, E. (1966). Ultrastructural variations in the anal papillae of *Aedes aegypti* (L.) at different environmental salinities. *J. Insect Physiol.* **12**, 429–439.

### Insect Malpighian Tubule

Baccetti, B., Mazzi, V., and Massimello, G. (1963). Ricerche istochimiche e al microscopio elettronico sui tubi Malpighiani di *Dacus oleae* Gmel. II. L, Adulto. *Redia* **48**, 47–68.

Byers, J. R. (1971). Metamorphosis of the perirectal Malpighian tubules in the mealworm *Tenebrio molitor* L. (Coleoptera, Tenebrionidae). II. Ultrastructure and role of autophagic vacuoles. *Cand. J. Zool.* **49**, 1185–1192.

Grinyer, I., and Musgrave, A. J. (1964). Microorganisms and mitochondria in the Malpighian tubules of *Sitophilus* (Coleoptera). *Can J. Microbiol.* **10**, 805–806.

Kessel, R. G. (1970). The permeability of dragonfly Malpighian tubule cells to protein using horseradish peroxidase as a tracer. *J. Cell Biol.* **47**, 299–303.

Maddrell, S. H. P. (1971). Fluid secretion by the Malpighian tubules of insects. *Phil. Trans. Roy. Soc.* London Ser. B **262**, 197–207.

Mazzi, V., and Baccetti, B. (1963). Ricerche istochimiche e al microscopio elettronico sui tubi malpighiani di *Dacus oleae* Gmel. I. La larva. *Z. Zellforsch. Mikrosk. Anat.* **59**, 47–70.

Messier, P.-E., and Sandborn, E. B. (1966). Mitochondries dans les microvillosités des tubes de Malpighi chez le grillon. *Rev. Can. Biol.* **25**, 217–219.

Meyer, G. F. (1957). Elektronmikroskopische Untersuchungen an den Malpighi-Gefässen verschiedener Insekten. *Z. Zellforsch. Mikrosk. Anat.* **47**, 18–28.

Taylor, H. H. (1971). Water and solute transport by the Malpighian tubules of the stick insect, *Carausius morosus*. The normal ultrastructure of the type I cells. *Z. Zellforsch. Mikrosk. Anat.* **118**, 333–368.

Tsubo, I., and Brandt, P. W. (1962). An electron microscopic study of the Malpighian tubules of the grasshopper, *Dissosteira carolina J. Ultrastruct. Res.* **6**, 28–35.

Wessing, A. (1965). Die funktion der Malpighischen Gefässe. *In* "Funktionelle und morphologische Organisation der Zelle," II. Sekretion und Exkretion., pp. 228–268. Springer-Verlag, Berlin.

Wessing, A., and Eichelberg, D. (1969). Elektronenmikroskopische Untersuchungen zur Lipid-Speicherung in den Nierentubuli von *Drosophila melanogaster. Z. Zellforsch. Mikrosk. Anat.* **94**, 129–146.

Wessing, A., and Eichelberg, D. (1969). Elektronenoptische Untersuchungen an den Nierentubuli (Malpighische Gefäße) von *Drosophila melanogaster.* I. Regionale Gliederung der Tubuli. *Z. Zellforsch. Mikrosk. Anat.* **101**, 285–322.

### Insect Rectum

Baccetti, B. (1962). Ricerche sull'ultrastruttura dell'intestino degli insetti. IV. Le papille rettali in un Ortottero adulto. *Redia* **47**, 105–118.

Gupta, B. L., and Berridge, M. J. (1966). A coat of repeating subunits on the cytoplasmic surface of the plasma membrane in the rectal papillae of the blowfly, *Calliphora erythrocephala* (Meig), studied *in situ* by electron microscopy. *J. Cell Biol.* **29**, 376–382.

Hopkins, C. R. (1966). The fine structural changes observed in the rectal papillae of the mosquito *Aedes aegypti* L. and their relation to the epithelial transport of water and inorganic ions. *J. Roy. Microsc. Soc.* **86**, 235–252.

Koefoed, B. M. (1971). Ultrastructure of the crytonephridial system in the meal worm *Tenebrio molitor. Z. Zellforsch. Mikrosk. Anat.* **116**, 487–501.

Noirot, C., and Noirot-Timothée, C. (1966). Revêtement de la membrane cytoplasmique et absorption des ions dans les papilles rectales d'un termite (Insecta, Isoptera). *C. R. Acad. Sci. Ser. D* **263**, 1099–1102.

Noirot-Timothée, C., and Noirot, C. (1966a). Attache de microtubules sur la membrane cellulaire dans le tube digestif des termites. *J. Microsc. Paris* **5**, 715–724.

Noirot-Timothée, C., and Noirot, C. (1966b). Liaison de mitochondries avec des zones d'adhésion intercellulaires. *J. Microsc. Paris* **6**, 87–90.

Saini, R. S. (1964). Histology and physiology of the cryptonephridial system of insects. *Trans. Roy. Ent. Soc. London* **116**, 347–392.

### Insect Goblet Cell

Akai, H. (1969). Ultrastructural localization of phosphatases in the midgut of the silkworm, *Bombyx mori. J. Insect Physiol.* **15**, 1623–1628.

## Amphibian Epidermis

Bracho, H., Erlij, D., and Martinez-Palomo, A. (1970). The site of the permeability barriers in frog skin epithelium. *J. Physiol.* 213, 50–51P.

Carasso, N., Favard, P., Jard, S., and Rajerison, R. M. (1971). The isolated frog skin epithelium. I. Preparation and general structure in different physiological states. *J. Microsc. Paris* 10, 315–330.

Erlij, D. (1971). Salt transport across isolated frog skin. *Phil. Trans. Roy. Soc. London Ser. B.* 262, 153–161.

Martinez-Palomo, A., Erlij, D., and Bracho, H. (1971). Localization of permeability barriers in the frog skin epithelium. *J. Cell Biol.* 50, 277–287.

Parakkal, P. F., and Matoltsy, A. G. (1964). A study of the fine structure of the epidermis of *Rana pipiens*. *J. Cell Biol.* 20, 85–94.

## Amphibian Urinary Bladder

Bartoszewicz, W., and Barrnett, R. J. (1964). Fine structural localization of nucleoside phosphatase activity in the urinary bladder of the toad. *J. Ultrastruct. Res.* 10, 599–609.

Choi, J. K. (1965). Electron microscopy of absorption of tracer materials by toad urinary bladder epithelium. *J. Cell Biol.* 25, 175–192.

Frazier, H. S. (1971). Sodium transport in the toad bladder: the functional organization of the granular cell: a review. *Circ. Res.* 28, Suppl. II, 14–20.

Hays, R. M., Singer, B., and Malamed, S. (1965). The effect of calcium withdrawal on the structure and function of the toad bladder. *J. Cell Biol.* 25, 195–208.

Jard, S., Bourguet, J., Carasso, N., and Favard, P. (1966). Action de divers fixateurs sur la perméabilité et l'ultrastructure de la vessie de grenouille. *J. Microsc. Paris* 5, 31–50.

Jard, S., Bouguet, J. Favard, P., and Carasso, N. (1971). The role of intercellular channels in the transepithelial transfer of water and sodium in the frog urinary bladder. *J. Membrane Biol.* 4, 124–147.

## Salt Gland

Cowan, F. B. M. (1971). The ultrastructure of the lachrymal 'salt' gland and the Harderian gland in the euryhaline *Malaclemys* and some closely related stenohaline emydines. *Can. J. Zool.* 49, 691–697.

Ellis, R. A., and Abel, J. H. (1964). Intercellular channels in the salt-secreting glands of marine turtles. *Science* 144, 1340–1342.

Ernst, S. A., and Philpott, C. W. (1970). Preservation of Na-K-activated and Mg-activated adenosine triphosphatase activities of avian salt gland and teleost gill with formaldehyde as fixative. *J. Histochem. Cytochem.* 18, 251–263.

Fawcett, D. W. (1962). Physiologically significant specializations of the cell surface. *Circulation* 26, 1105–1125.

Komnick, H. (1963). Elektronenmikroskopische Untersuchungen zur funktionellen Morphologie des Ionentransportes in der Salzdrüse von *Larus argentatus*. II. Teil: Funktionelle Morphologie der Blutgefässe. *Protoplasma* 56, 385–419.

Komnick, H. (1964). Elektronenmikroskopische Untersuchungen zur funktionellen Morphologie des Ionentransportes in der Salzdrüse von *Larus argentatus*. IV. Teil: Funktionelle Morphologie der Epithelzellen des Sammelkanals. *Protoplasma* 58, 96–127.

Komnick, H., and Kniprath, E. (1970). Morphometric studies on the herring gull salt gland. *Cytobiologie* 1, 228–247.

Komnick, H., and Komnick, U. (1963). Elektronenmikroskopische Untersuchungen zur funtionellen Morphologie des Ionentransportes in der Salzdrüse von *Larus argentatus*. V. Teil: Experimenteller Nachweis der Transportwege. *Z. Zellforsch. Mikrosk. Anat.* 60, 163–203.

## Fish Gills

Copeland, D. E., and Dalton, A. J. (1959). An association between mitochondria and the endoplasmic reticulum in cells of the pseudobranch gland of a teleost. *J. Biophys. Biochem. Cytol.* 5, 393–396.

Doyle, W. L., and Gorecki, D. (1961). The so-called chloride cells of the fish gill. *Physiol. Zool.* 34, 81–85.

Harb, J. M., and Copeland, D. E. (1969). Fine structure of the pseudobranch of the flounder *Paralichthys lethostigma*. A description of a chloride-type cell and pseudobranch-type cell. *Z. Zellforsch. Mikrosk. Anat.* 101, 167–174.

Newstead, J. D. (1971). Observations on the relationship between "chloride-type" and "pseudobranch-type" cells in the gills of a fish, *Oligocottus maculosus*. *Z. Zellforsch. Mikrosk. Anat.* 116, 1–6.

Philpott, C. W. (1968). Functional implications of the cell surface: the plasmalemma and surface-associated polyanions. *In* "Cystic Fibrosis" (R. Porter and M. O'Connor, eds.), pp. 109–116, *Ciba Found. Study Group No. 32*, Churchill, London.

Öberg, K. E. (1967). The reversibility of the respiratory inhibition in gills and the ultrastructural changes in chloride cells from the rotenone-poisoned marine teleost, *Gadus callarias* L. *Exp. Cell Res.* 45, 590–602.

Petřík, P. (1968). The demonstration of chloride ions in the "chloride cells" of the gills of eels (*Anguilla anguilla* L.) adapted to sea water. *Z. Zellforsch. Mikrosk. Anat.* 92, 422–427.

Petřík, P., and Bucher, O. (1969). A propos des "chloride cells" dans l'épithélium des lamelles branchiales du poisson rouge. *Z. Zellforsch. Mikrosk. Anat.* 96, 66–74.

## Mammalian Kidney

Anderson, W. A. (1967). The fine structure of compensatory growth in the rat kidney after unilateral nephrectomy. *Amer. J. Anat.* 121, 217–248.

Bulger, R. E., and Trump, B. F. (1966). Fine structure of the rat renal papilla. *Amer. J. Anat.* 118, 685–722.

Bulger, R. E., and Trump, B. F. (1968). Renal morphology of the English Sole. *Amer. J. Anat.* **123**, 195–226.

Bulger, R. E., and Trump, B. F. (1969). A mechanism for rapid transport of colloidal particles by flounder renal epithelium. *J. Morphol.* **127**, 205–224.

Bulger, R. E., and Trump, B. F. (1969). Ultrastructure of granulated arteriolar cells (juxtaglomerular cells) in kidney of a fresh and a salt water teleost. *Amer. J. Anat.* **124**, 77–88.

Bulger, R. E., Griffith, L. D., and Trump, B. F. (1966). Endoplasmic reticulum in rat renal interstitial cells: Molecular rearrangement after water deprivation. *Science* **151**, 83–86.

Ericsson, J. L. E., and Trump, B. F. (1966). Electron microscopic studies of the epithelium of the proximal tubule of the rat kidney. III. Microbodies, multivesicular bodies and Golgi apparatus. *Lab. Invest.* **15**, 1610–1633.

Ganote, C. E., Grantham, J. J., Moses, H. L., Burg, M. B., and Orloff, J. (1968). Ultrastructural studies of vasopressin effect on isolated perfused renal collecting tubules of the rabbit. *J. Cell Biol.* **36**, 355–367.

Giebisch, G. (1969). Functional organization of proximal and distal tubular electrolyte transport. *Nephron* **6**, 260–281.

Graham, R. C., and Karnovsky, M. J. (1966). The early stages of absorption of injected horseradish peroxidase in the proximal tubules of mouse kidney: Ultrastructural cytochemistry by a new technique. *J. Histochem. Cytochem.* **14**, 291–302.

Groniowski, J., Biczyskowa, W., and Walski, M. (1969). Electron microscope studies on the surface coat of the nephron. *J. Cell Biol.* **40**, 585–601.

Heller, J., and Rossmann, P. (1969). The significance of the intercellular spaces in proximal tubular reabsorption in the rat. *Physiol. Bohemoslov.* **18**, 227–232.

Latta, H., and Maunsbach, A. B. (1962). Relations of the centrolobular region of the glomerulus to the juxtaglomerular apparatus. *J. Ultrastruct. Res.* **6**, 562–578.

de Martino, C., and Zamboni, L. (1966). A morphologic study of the mesonephros of the human embryo. *J. Ultrastruct. Res.* **16**, 399–427.

Pease, D. C. (1955). Electron microscopy of the tubular cells of the kidney cortex. *Anat. Rec.* **121**, 723–743.

Sabour, M. S., MacDonald, M. K., Lambie, A. T., and Robson, J. S. (1964). The electron microscopic appearance of the kidney in hydrated and dehydrated rats. *Quart. J. Exp. Physiol.* **49**, 162–170.

Tisher, C. C., Finkel, R. M., Rosen, S., and Kendig, E. M. (1968). Renal microbodies in the Rhesus monkey. *Lab. Invest.* **19**, 1–6.

Trump, B. F. (1961). An electron microscope study of the uptake, transport, and storage of colloidal material by the cells of the vertebrate nephron. *J. Ultrastruct. Res.* **5**, 291–310.

Waugh, D., Prentice, R. S. A., and Yadav, D. (1967). The structure of the proximal tubule: A morphological study of basement membrane cristae and their relationships in the renal tubule of the rat. *Amer. J. Anat.* **121**, 775–786.

## Mammalian Salivary Glands

Amsterdam, A., Ohad, I., and Schramm, M. (1969). Dynamic changes in the ultrastructure of the acinar cell of the rat parotid gland during the secretory cycle. *J. Cell Biol.* **41**, 753–773.

Bogart, B. I. (1970). The effect of aging on the rat submandibular gland: An ultrastructural, cytochemical and biochemical study. *J. Morphol.* **130**, 337–351.

Caramia, F. (1966a). Ultrastructure of mouse submaxillary gland. I. Sexual differences. *J. Ultrastruct. Res.* **16**, 505–523.

Caramia, F. (1966b). Ultrastructure of mouse submaxillary gland. II. Effect of castration in the male. *J. Ultrastruct. Res.* **16**, 524–536.

Cowley, L. H., and Shackleford, J. M. (1970). An ultrastructural study of the submandibular glands of the squirrel monkey, *Saimiri sciureus*. *J. Morphol.* **132**, 117–135.

Ferner, H., and Gansler, H. (1961). Elektronenmikroskopische Untersuchungen an der Glandula Submandibularis und Parotis des Menschen. *Z. Zellforsch. Mikrosk. Anat.* **55**, 148–178.

Flatland, R. F., Schneyer, L. H., and Schneyer, C. A. (1968). Amylase activity of acinar cells separated from guinea pig submaxillary gland. *Proc. Soc. Exp. Biol. Med.* **131**, 243–246.

Kagayama, M. (1971). The fine structure of the monkey submandibular gland with a special reference to intra-acinar nerve endings. *Amer. J. Anat.* **131**, 185–196.

Leeson, C. R. (1967). Structure of salivary glands. *In* "Handbook of Physiology," Sect. 6, Vol. II, pp. 463–495. Amer. Physiol. Soc., Wash.

Luzzatto, A. C., Procicchiani, G., and Rosati, G. (1968). Rat submaxillary gland: An electron microscope study of the secretory granules of the acinus. *J. Ultrastruct. Res.* **22**, 185–194.

Parks, H. F. (1962). Morphological study of the extrusion of secretory materials by the parotid glands of mouse and rat. *J. Ultrastruct. Res.* **6**, 449–465.

Radley, J. M. (1969). Ultrastructural changes in the rat submaxillary gland following isoprenaline. *Z. Zellforsch. Mikrosk. Anat.* **97**, 196–211.

Scott, B. L., and Pease, D. C. (1959). Electron microscopy of the salivary and lacrimal glands of the rat. *Amer. J. Anat.* **104**, 115–161.

Shackleford, J. M., and Schneyer, C. A. (1964). Structural and functional aspects of rodent salivary glands including two desert species. *Amer. J. Anat.* **115**, 279–307.

Shackleford, J. M., and Schneyer, L. H. (1971). Ultrastructural aspects of the main excretory duct of rat submandibular gland. *Anat. Rec.* **169**, 679–696.

Shackleford, J. M., and Wilborn, W. H. (1970a). Ultrastructural aspects of cat submandibular glands. *J. Morphol.* **131**, 253–276.

Shackleford, J. M., and Wilborn, W. H. (1970b). Ultrastructural aspects of calf submandibular glands. *Amer. J. Anat.* **127**, 259–280.

Simson, J. V. (1969). Discharge and restitution of secretory material in the rat parotid gland in response to isoproterenol. *Z. Zellforsch. Mikrosk. Anat.* **101**, 175–191.

Strum, J. M., and Karnovsky, M. J. (1970). Ultrastructural localization of peroxidase in submaxillary acinar cells. *J. Ultrastruct. Res.* **31**, 323–336.

Takahama, M., and Barka, T. (1967). Electron microscopic alterations of submaxillary gland produced by isoproterenol. *J. Ultrastruct. Res.* **17**, 452–474.

Tandler, B., and Ross, L. L. (1969). Observations of nerve terminals in human labial salivary glands. *J. Cell Biol.* **42**, 339–343.

Tandler, B., Denning, C. R., Mandel, I. D., and Kutscher, A. H. (1969). Ultrastructure of human labial salivary glands. I. Acinar secretory cells. *J. Morphol.* **127**, 383–408.

Tandler, B., Denning, C. R., Mandel, I. D., and Kutscher, A. H. (1970). Ultrastructure of human labial salivary glands. III. Myoepithelium and ducts. *J. Morphol.* **130**, 227–246.

Tapp, R. L. (1967). The ultrastructure of watery vacuoles in the submandibular gland of the rat. *J. Roy. Microsc. Soc.* **88**, 1–11.

Wilborn, W. H., and Schneyer, C. A. (1970). Ultrastructural changes of rat parotid glands induced by a diet of liquid Metrecal. *Z. Zellforsch. Mikrosk. Anat.* **103**, 1–11.

## Insect Salivary Gland

Bland, K. P., and House, C. R. (1971). Function of the salivary glands of the cockroach, *Nauphoeta cinerea*. *J. Insect Physiol.* **17**, 2069–2084.

Lane, N. J., Ashburner, M., and Carter, Y. R. (1972). Puffs and salivary gland function: The fine structure of the larval and prepupal salivary glands of *Drosophila melanogaster*. Wilhelm Roux' Archiv. (in press).

MacGregor, H. C., and Mackie, J. B. (1967). Fine structure of the cytoplasm in salivary glands of *Simulium*. *J. Cell Sci.* **2**, 137–144.

Moericke, V., and Wohlfarth-Bottermann, K. E. (1960a). Zur Funktionellen Morphologie der Speicheldrüsen von Homopteren. I. Mitteilung: Die Hauptzellen der Hauptdrüse von *Myzus persicae* (Sulz.), Aphididae. *Z. Zellforsch. Mikrosk. Anat.* **51**, 157–184.

Moericke, V., and Wohlfarth-Bottermann, K. E. (1960b). Zur Funktionellen Morphologie der Speicheldrüsen von Homopteren. IV. Mitteilung: Die Ausführgänge der Speicheldrüsen von *Myzus persicae* (Sulz.), Aphididae. *Z. Zellforsch. Mikrosk. Anat.* **53**, 25–49.

Phillips, D. M., and Swift, H. (1965). Cytoplasmic fine structure of *Sciara* salivary glands. I. Secretion. *J. Cell Biol.* **27**, 395–409.

Poels, C. L. M., de Loof, A., and Berendes, H. D. (1971). Functional and structural changes in *Drosophila* salivary gland cells triggered by ecdysterone. *J. Insect Physiol.* **17**, 1717–1729.

Schin, K. S., and Clever, U. (1968). Ultrastructural and cytochemical studies of salivary gland regression in *Chironomus tentans*. *Z. Zellforsch. Mikrosk. Anat.* **86**, 262–279.

Welsch, U., Wächtler, K., and Rühm, W. (1968). Die Feinstruktur der Speicheldrüse von *Boophthora*

*erythrocephala* de GEER (Simuliidae, Diptera) vor und nach der Blutaufnahme. *Z. Zellforsch. Mikrosk. Anat.* **88**, 304–352.

Whitehead, A. T. (1972). The innervation of the salivary gland in the cockroach: Light and electron microscopy observations. *J. Morphol.* **135**, 483–506.

Wiener, J., Spiro, D., and Loewenstein, W. R. (1964). Studies on an epithelial (gland) cell junction. II. Surface structure. *J. Cell Biol.* **22**, 587–598.

Wohlfarth-Bottermann, K. E., and Moericke, V. (1960). Zur Funktionellen Morphologie der Speicheldrüsen von Homopteren. III. Mitteilung: Die Nebendrüse von *Myzus persicae* (Sulz.), Aphididae. *Z. Zellforsch. Mikrosk. Anat.* **52**, 346–361.

## Parietal Cell of Vertebrate Stomach

Forte, G. M., Limlomwongse, L., and Forte, J. G. (1969). The development of intracellular membranes concomitant with the appearance of HCl secretion in oxyntic cells of the metamorphosing bullfrog tadpole. *J. Cell Sci.* **4**, 709–727.

Gusek, W. (1961). Zur Ultramikroskopischen Cytologie der Belegzellen in der Magenschleimhaut des Menschen. *Z. Zellforsch. Mikrosk. Anat.* **55**, 790–809.

Hally, A. D. (1959). The fine structure of the gastric parietal cell in the mouse. *J. Anat.* **93**, 217–225.

Hayward, A. F. (1967). The ultrastructure of developing gastric parietal cells in the foetal rabbit. *J. Anat.* **101**, 69–81.

Helander, H. F. (1964a). Ultrastructure of secretory cells in the pyloric gland area of the mouse gastric mucosa. *J. Ultrastruct. Res.* **10**, 145–159.

Helander, H. F. (1964b). Ultrastructure of gastric fundus glands of refed mice. *J. Ultrastruct. Res.* **10**, 160–175.

Helander, H., and Ekholm, R. (1959). Ultrastructure of epithelial cells in the fundus glands of the mouse gastric mucosa. *J. Ultrastruct. Res.* **3**, 74–83.

Ito, S. (1967). Anatomic structure of the gastric mucosa. *In* "Handbook of Physiology," Sect. 6, Vol. II, pp. 705–741. Amer. Physiol. Soc., Washington.

Kurosumi, F., Shibasaki, S., Uchida, G., and Tanaka, Y. (1958). Electron microscope studies on the gastric mucosa of normal rats. *Arch. Histol. Jap.* **15**, 587.

Lawn, A. M. (1960). Observations on the fine structure of the gastric parietal cell of the rat. *J. Biophys. Biochem. Cytol.* **7**, 161–166.

Sedar, A. W., and Forte, J. G. (1964). Effects of calcium depletion on the junctional complex between oxyntic cells of gastric glands. *J. Cell Biol.* **22**, 173–188.

Toner, P. G. (1963). The fine structure of resting and active cells in the submucosal glands of the fowl proventriculus. *J. Anat.* **97**, 575–583.

## Vertebrate Liver

Wisse, E. (1970). An electron microscopic study of the fenestrated endothelial lining of rat liver sinusoids. *J. Ultrastruct. Res.* **31**, 125–150.

## Vertebrate Gallbladder

Bader, G. (1965). Die submikroskopische Struktur des Gallenblasenepithels und seiner Regeneration. I. Mitteilung: Karpfen (*Cyprinus carpio*, L.) und Frosch (*Rana esculenta*, L.). Z. Mikrosk. Anat. Forsch. 74, 92–107.

Hayward, A. F. (1962). Electron microscopic observations on absorption in the epithelium of the guinea pig gall bladder. Z. Zellforsch. Mikrosk. Anat. 56, 197–202.

Yamada, E. (1955). The fine structure of the gall bladder epithelium of the mouse. J. Biophys. Biochem. Cytol. 1, 445–458.

Yamada, K. (1968). Some observations on the fine structure of light and dark cells in the gall bladder epithelium of the mouse. Z. Zellforsch. Mikrosk. Anat. 84, 463–472.

Yamada, K. (1969). Fine structure of rodent common bile duct epithelium. J. Anat. 105, 511–523.

## Vertebrate Exocrine Pancreas

Legg, P. G. (1968). Electron microscopic studies on compound tubular bodies in acinar cells of cat pancreas. J. Anat. 103, 359–370.

## Vertebrate Small Intestine

Bergener, M. (1962). Die Feinstruktur des Dünndarmepithels während der Physiologischen Milchresorption beim Jungen Goldhamster. Z. Zellforsch. Mikrosk. Anat. 57, 428–474.

Bockman, D. E., and Winborn, W. B. (1966). Light and electron microscopy of intestinal ferritin absorption. Observations in sensitized and non-sensitized hamsters (*Mesocricetus auratus*). Anat. Rec. 155, 603–622.

Bonneville, M. A. (1963). Fine structural changes in the intestinal epithelium of the bullfrog during metamorphosis. J. Cell Biol. 18, 579–597.

Boyd, C. A. R., and Parsons, D. S. (1969). The fine structure of the microvilli of isolated brush borders of intestinal epithelial cells. J. Cell Biol. 41, 646–651.

Brown, A. L. (1962). Microvilli of the human jejunal epithelial cell. J. Cell Biol. 12, 623–627.

Dermer, G. B. (1967a), Ultrastructural changes in the microvillous plasma membrane during lipid absorption and the form of absorbed lipid: An *in vitro* study. J. Ultrastruct. Res. 20, 51–71.

Dermer, G. B. (1967b). Ultrastructural changes in the microvillus plasma membrane during lipid absorption and the form of absorbed lipid: An *in vitro* study at 37°C. J. Ultrastruct. Res. 20, 311–320.

Dermer, G. B. (1967c). Ultrastructural changes in the microvillous plasma membrane during lipid absorption: An *in vitro* study at 0°C. J. Ultrastruct. Res. 21, 1–8.

Horstmann, E. (1966). Über das endothel der zottenkapillaren im Dünndarm des Meerschweinchens und des Menschen. Z. Zellforsch. Mikrosk. Anat. 72, 364–369.

Hourdry, J. (1969). Ultrastructure de l'épithélium intestinal larvaire chez un Amphibien Anoure, *Alytes obstetricans* Laur. Z. Zellforsch. Mikrosk. Anat. 94, 574–592.

Iwai, T. (1968). Fine structure and absorption patterns of intestinal epithelial cells in rainbow trout alevins. Z. Zellforsch. Mikrosk. Anat. 91, 366–379.

Kraehenbuhl, J.-P., and Campiche, M. A. (1969). Early stages of intestinal absorption of specific antibodies in the newborn. An ultrastructural, cytochemical, and immunological study in the pig, rat, and rabbit. J. Cell Biol. 42, 345–365.

Kraehenbuhl, J.-P., Gloor, E., and Blanc, B. (1966). Morphologie comparée de la muqueuse intestinale de deux espèces animales aux possibilités d'absorption protéique néonatale différentes. Z. Zellforsch. Mikrosk. Anat. 70, 209–219.

McNabb, J. D., and Sandborn, E. (1964). Filaments in the microvillous border of intestinal cells. J. Cell Biol. 22, 701–704.

Millington, P. F., and Finean, J. B. (1962). Electron microscope studies of the structure of the microvilli on principal epithelial cells of rat jejunum after treatment in hypo- and hypertonic saline. J. Cell Biol. 14, 125–139.

Mukherjee, T. M., and Williams, A. W. (1967). A comparative study of the ultrastructure of microvilli in the epithelium of small and large intestine of mice. J. Cell Biol. 34, 447–461.

Ruska, C. (1960). Die Zellstrukturen des Dünndarmepithels in ihrer Abhängigkeit von der Physikalisch-Chemischen Beschaffenheit des Darminhalts. I. Wasser and Natriumchlorid. Z. Zellforsch. Mikrosk. Anat. 52, 748–777.

Schmidt, W. (1961). Elektronenmikroskopische Untersuchung des Intrazellulären Stofftransportes in der Dünndarmepithelzelle nach Markierung mit Myofer. Z. Zellforsch. Mikrosk. Anat. 54, 803–806.

Sibalin, M., and Björkman, N. (1966). On the fine structure and absorptive function of the porcine jejunal villi during the early suckling period. Exp. Cell Res. 44, 165–174.

Silva, D. G. (1966). The fine structure of multivesicular cells with large microvilli in the epithelium of the mouse colon. J. Ultrastruct. Res. 16, 693–705.

Sjöstrand, F. S. (1963). The ultrastructure of the plasma membrane of columnar epithelium cells of the mouse intestine. J. Ultrastruct. Res. 8, 517–541.

Strauss, E. W. (1963). The absorption of fat by intestine of golden hamster *in vitro*. J. Cell Biol. 17, 597–607.

Tomasini, J. T., and Dobbins, W. O. (1970). Intestinal mucosal morphology during water and electrolyte absorption. A light and electron microscopic study. Amer. J. Dig. Dis. 15, 226–238.

Wissig, S. L., and Graney, D. O. (1968). Membrane modifications in the apical endocytic complex of ileal epithelial cells. J. Cell Biol. 39, 564–579.

Yamamoto, T. (1966). An electron microscope study of the columnar epithelial cell in the intestine of fresh water teleosts: goldfish (*Carassius auratus*) and rainbow trout (*Salmo irideus*). Z. Zellforsch. Mikrosk. Anat. 72, 66–87.

## Insect Midgut

(For additional references, see Insect Goblet Cell, p. 83)

Hecker, H., Freyvogel, T. A., Briegel, H., and Steiger, R. (1971). Ultrastructural differentiation of the midgut epithelium in female *Aedes aegypti* (L.) (Insecta, Diptera) imagines. *Acta Trop.* **28,** 80–104.

Jones, J. C., and Zeve, V. H. (1968). The fine structure of the gastric caeca of *Aedes aegypti* larvae. *J. Insect Physiol.* **14,** 1567–1575.

Smith, D. S., Compher, K., Janners, M., Lipton, C., and Wittle, L. W. (1969). Cellular organization and ferritin uptake in the midgut epithelium of a moth, *Ephestia kühniella. J. Morphol.* **127,** 41–72.

Waterhouse, D. F., and Wright, M. (1960). The fine structure of the mosaic midgut epithelium of blowfly larvae. *J. Insect Physiol.* **5,** 230–239.

## Rumen

Henrikson, R. C. (1970). Developmental changes in the structure of perinatal ruminal epithelium: basal infoldings, glycogen, and glycocalyx. *Z. Zellforsch. Mikrosk. Anat.* **109,** 15–19.

## Choroid Plexus

Carpenter, S. J. (1966). An electron microscopic study of the choroid plexuses in *Necturus maculosus. J. Comp. Neurol.* **127,** 413–433.

Cserr, H. F. (1971). Physiology of the choroid plexus. *Physiol. Rev.* **51,** 273–311.

Becker, N. H., and Sutton, C. H. (1963). Pathologic features of the choroid plexus. I. Cytochemical effects of hypervitaminosis A. *Amer. J. Pathol.* **43,** 1017–1030.

Becker, N. H., Novikoff, A. B., and Zimmerman, H. M. (1967). Fine structure observations of the uptake of intravenously injected peroxidase by the rat choroid plexus. *J. Histochem. Cytochem.* **15,** 160–165.

Dohrmann, G. J. (1970a). The choroid plexus: A historical review. *Brain Res.* **18,** 197–218.

Dohrmann, G. J. (1970b). Dark and light epithelial cells in the choroid plexus of mammals. *J. Ultrastruct. Res.* **32,** 268–273.

Dohrmann, G. J., and Herdson, P. B. (1969). Lobated nuclei in epithelial cells of the choroid plexus of young mice. *J. Ultrastruct. Res.* **29,** 218–223.

Meller, K., and Wagner, H. H. (1968a). Die Feinstruktur des Plexus chorioideus in Gewebekulturen. *Z. Zellforsch. Mikrosk. Anat.* **86,** 98–110.

Meller, K., and Wagner, H. H. (1968b). Vergleichende elektronen-mikroskopische Untersuchungen des Plexus chorioideus der Maus in vivo and in vitro. *Z. Zellforsch. Mikrosk. Anat.* **91,** 507–518.

Meller, K., and Wechsler, W. (1965). Elektronenmikroskopische Untersuchung der Entwicklung der Telencephalen Plexus Chorioides des Huhnes. *Z. Zellforsch. Mikrosk. Anat.* **65,** 420–444.

Paul, E. (1970). Lipidkugeln im Plexusepithel von *Rana temporaria* L. *Z. Zellforsch. Mikrosk. Anat.* **106,** 539–549.

Pontenagel, M. (1962). Elektronenmikroskopische Untersuchungen am Ependym der Plexus chorioidei bei *Rana esculenta* und *Rana fusca* (Roesel). *Z. Mikrosk. Anat. Forsch.* **68,** 371–392.

Rodriguez, E. M., and Heller, H. (1970). Antidiuretic activity and ultrastructure of the toad choroid plexus. *J. Endocrinol.* **46,** 83–91.

Weindl, A., Schinko, I., Wetzstein, R., and Herz, A. (1969). Die sphärischen Lipidkörper im Epithel des Plexus chorioideus beim Kaninchen. *Z. Zellforsch. Mikrosk. Anat.* **100,** 300–315.

Wislocki, G. B., and Ladman, A. J. (1958). The fine structure of the mammalian choroid plexus. *In* "The Cerebrospinal fluid," pp. 55–79. Churchill, London.

## Stria Vascularis of Vertebrate Inner Ear

Nakai, Y., and Hilding, D. A. (1966). Electron microscopic studies of adenosine triphosphatase activity in the stria vascularis and spiral ligament. *Acta Oto-Laryngol.* **62,** 411–428.

## Mammary Gland

Bousquet, M., Fléchon, J. E., and Denamur, R. (1969). Aspects ultrastructuraux de la glande mammaire de lapine pendant la lactogénèse. *Z. Zellforsch. Mikrosk. Anat.* **96,** 418–436.

Girardie, J. (1968). Histo-cytomorphologie de la glande mammaire de la souris C3H et de trois autres rongeurs. *Z. Zellforsch. Mikrosk. Anat.* **87,** 478–503.

Helminen, H. J., and Ericsson, J. L. E. (1968a). Studies on mammary gland involution. II. Ultrastructural evidence for auto- and heterophagocytosis. *J. Ultrastruct. Res.* **25,** 214–227.

Helminen, H. J., and Ericsson, J. L. E. (1968b). Studies on mammary gland involution. III. Alterations outside auto- and heterophagocytic pathways for cytoplasmic degradation. *J. Ultrastruct. Res.* **25,** 228–239.

Helminen, H. J., Ericsson, J. L. E., and Orrenius, S. (1968). Studies on mammary gland involution. IV. Histochemical and biochemical observations on alterations in lysosomes and lysosomal enzymes. *J. Ultrastruct. Res.* **25,** 240–252.

Hollmann, K. H. (1966). Sur des aspects particuliers des protéines élaborées dans la glande mammaire. Étude au microscope électronique chez la lapine en lactation. *Z. Zellforsch. Mikrosk. Anat.* **69,** 395–402.

Hollmann, K. H., and Verley, J. M. (1966). Individualisation au microscope optique des grains de proteine secretes par la glande mammaire. *Z. Zellforsch. Mikrosk. Anat.* **75,** 601–604.

Hollman, K. H., and Verley, J. M. (1967). La régression de la glande mammaire à l'arrêt de la lactation. II. Étude au microscope électronique. *Z. Zellforsch. Mikrosk. Anat.* **82,** 222–238.

Sekhri, K. K., Pitelka, D. R., and DeOme, K. B. (1967a). Studies of mouse mammary glands. I. Cytomorphology of the normal mammary gland. *J. Nat. Cancer Inst.* **39**, 459–490.

Sekhri, K. K., Pitelka, D. R., and DeOme, K. B. (1967b). Studies of mouse mammary glands. II. Cytomorphology of mammary transplants in inguinal fat pads, nipple-excised host glands, and whole mammary-gland transplants. *J. Nat. Cancer Inst.* **39**, 491–527.

Stockinger, L., and Zarzicki, J. (1962). Elektronenmikroskopische Untersuchungen der Milchdrüse des Laktierenden Meerschweinchens mit Berücksichtigung des Saugaktes. *Z. Zellforsch. Mikrosk. Anat.* **57**, 106–123.

Toker, C. (1967). Observations on the ultrastructure of a mammary ductule. *J. Ultrastruct. Res.* **21**, 9–25.

Wellings, S. R., and DeOme, K. B. (1961). Milk protein droplet formation in the Golgi apparatus of the C3H/Crgl mouse mammary epithelial cells. *J. Biophys. Biochem. Cytol.* **9**, 479–485.

Wellings, S. R., and DeOme, K. B. (1963). Electron microscopy of milk secretion in the mammary gland of the C3H/Crgl mouse. III. Cytomorphology of the involuting gland. *J. Nat. Cancer Inst.* **30**, 241–267.

Wellings, S. R., DeOme, K. B., and Pitelka, D. R. (1960). Electron microscopy of milk secretion in the mammary gland of the C3H/Crgl mouse. I. Cytomorphology of the prelactating and the lactating gland. *J. Nat. Cancer Inst.* **25**, 393–421.

## Sweat Glands

Biempica, L., and Montes, L. F. (1965). Secretory epithelium of the large axillary sweat glands. *Amer. J. Anat.* **117**, 47–72.

Brusilow, S. W., Ikai, K., and Gordes, E. (1968). Comparative physiological aspects of solute secretion by the eccrine sweat gland of the rat. *Proc. Soc. Exp. Biol. Med.* **129**, 731–732.

Hashimoto, K., Gross, B. G., and Lever, W. E. (1966). Electron microscopic study of the human adult eccrine sweat gland. *J. Invest. Dermatol.* **46**, 172–185.

Lee, M. M. C. (1960). Histology and histochemistry of human eccrine sweat glands, with special reference to their defence mechanisms. *Anat. Rec.* **136**, 97–106.

Matsuzawa, T., and Kurosumi, K. (1963). The ultrastructure, morphogenesis, and histochemistry of the sweat glands in the rat foot pads as revealed by electron microscopy. *J. Electron Microsc.* **12**, 175–191.

Munger, B. L. (1961). The ultrastructure and histophysiology of human eccrine sweat glands. *J. Biophys. Biochem. Cytol.* **11**, 385–402.

Munger, B. L., and Brusilow, S. W. (1961). An electron microscopic study of eccrine sweat glands of the cat foot and toe pads—evidence for ductal reabsorption in the human. *J. Biophys. Biochem. Cytol.* **11**, 403–417.

Munger, B. L., and Brusilow, S. W. (1971). The histophysiology of rat plantar sweat glands. *Anat. Rec.* **169**, 1–22.

Ochi, J. (1968). Elektronenmikroskopischer Nachweis der Natriumionen in den Schweissdrüsen der Rattenfusssohl. *Histochemie* **14**, 300–307.

Quatrale, R. P., and Laden, K. (1968). Solute and water secretion by the eccrine sweat glands of the rat. *J. Invest. Dermatol.* **51**, 502–504.

Terzakis, J. A. (1964). The ultrastructure of monkey eccrine sweat glands. *Z. Zellforsch. Mikrosk. Anat.* **64**, 493–509.

Weiner, J. S., and Hellmann, K. (1960). The sweat glands. *Biol. Rev.* **35**, 141–186.

## Ciliary Epithelium of Vertebrate Eye

Brini, A., and Porte, A. (1959). Etude du corps ciliaire au microscope électronique. *Bull. Soc. Fr. Ophthalmol.* **72**, 56–72.

Fine, B. S., and Zimmerman, L. E. (1963). Light and electron microscopic observations on the ciliary epithelium in man and rhesus monkey: with particular reference to the base of the vitreous body. *Invest. Ophthalmol.* **2**, 105–137.

Kaye, G. I., and Pappas, G. D. (1965). Studies on the ciliary epithelium and zonule. III. The fine structure of the rabbit ciliary epithelium in relation to the localization of ATPase activity. *J. Microsc. Paris* **4**, 497–508.

Holmberg, A. (1959). Ultrastructure of the ciliary epithelium. *AMA Arch. Ophthalmol.* **62**, 935–948.

Pappas, G. D., and Smelser, G. K. (1958a). The fine structure of the ciliary epithelium in relation to aqueous humor secretion. In "The Structure of the Eye" (G. K. Smelser, ed.), pp. 453–467. Academic Press, New York.

Pappas, G. D., and Smelser, G. K. (1958b). Studies on the ciliary epithelium and the zonule: I. Electron microscope observations on changes induced by alteration of normal aqueous humor formation in the rabbit. *Amer. J. Opthalmol.* **46**, 299–318.

Pappas, G. D., Smelser, G. K., and Brandt, P. W. (1959). Studies on the ciliary epithelium and the zonule. II. Electron and fluorescence microscope observations on the function of membrane elaborations. *Arch. Ophthalmol.* **62**, 959–965.

Pappas, G. D. (1959). Ultrastructure of the ciliary epithelium and its relationship to aqueous secretion. In "Glaucoma, *Trans 4th Conf.*" (F. W. Newell, ed.) pp. 141–178. Josiah Macy Jr. Foundations, New York.

Tormey, J. McD. (1965). Artifactual localization of ferritin in the ciliary epithelium in vitro. *J. Cell Biol.* **25**, 1–7.

Tormey, J. McD. (1966). The ciliary epithelium: An attempt to correlate structure and function. Symposium: Contribution of Electron Microscopy to the understanding of the production and outflow of aqueous humor. *Trans. Amer. Acad. Opthalmol. Otolaryngol*, Sept.–Oct. 1966.